UN Robust Peacekeeping

UN Robust Peacekeeping

Civilian Protection in Violent Civil Wars

Kofi Nsia-Pepra

palgrave
macmillan

UN ROBUST PEACEKEEPING

First published in 2014 by
PALGRAVE MACMILLAN®
in the United States—a division of St. Martin's Press LLC,
175 Fifth Avenue, New York, NY 10010.

Where this book is distributed in the UK, Europe and the rest of the World, this is by Palgrave Macmillan, a division of Macmillan Publishers Limited, registered in England, company number 785998, of Houndmills, Basingstoke, Hampshire RG21 6XS.

Palgrave Macmillan is the global academic imprint of the above companies and has companies and representatives throughout the world.

Palgrave® and Macmillan® are registered trademarks in the United States, the United Kingdom, Europe and other countries.

ISBN: 978–1–137–39785–0

Library of Congress Cataloging-in-Publication Data

Nsia-Pepra, Kofi.
United Nations robust peacekeeping : civilian protection in violent civil wars / by Kofi Nsia-Pepra.
pages cm.
ISBN 978–1–137–39785–0 (hardcover : alk. paper)
1. United Nations—Peacekeeping forces. 2. Peace-building—International cooperation. 3. Security, International—International cooperation. I. Title.
JZ6374.N75 2014
341.5′84—dc23 2014006495

A catalogue record of the book is available from the British Library.

Design by Integra Software Services

First edition: August 2014

10 9 8 7 6 5 4 3 2 1

This book is dedicated to my parents, children, brothers and sisters, all my loved ones and victims of human rights violations

Contents

Tables

Figures

Preface

The book is motivated by my personal experiences of human rights violations as a UN peacekeeper with United Nations Assisted Mission in Rwanda (UNAMIR) and Ghana Air Force Detachment Commander with ECOMOG in the Sierra Leone war. My experiences in both missions also equipped me with practical evidences of the failures of traditional peacekeeping in Rwanda and the efficacy of robust peacekeeping in protecting vulnerable civilians from attacks by spoilers in violent civil wars. As we drove over the beautiful mountains in the Rwamatamu *commune*, Kibuye prefecture, Rwanda, almost every stop showed us new horrors: torched churches filled with charred corpses of people who had prayed for mercy in a holy place but rather betrayed by presumed *Saviors* and *Men of God* who wickedly handed them over to the *Interahamwe* to be massacred in the *House of God*, . . . and everywhere bodies, decaying or picked clean to the bones. This was human tragedy and cost, the *reduction ut horrendous*, that came from traditional peacekeeping handed to Rwanda during genocide despite Rwandese demand for and global agreement on a *robust mission* during the Arusha peace agreement. UNAMIR, commanded by Canadian General *Roméo* Dellaire, ostensibly sent there to abate the severity and intensity of the conflict and protect Tutsis and moderate Hutu's from radical Hutu power militias, helplessly stood by and did nothing as approximately 900,000 people were butchered in what has been described as Africa's worst human tragedy. Dellaire had been desperate to use force to stop the genocide, but the United Nations repeatedly ordered him to do nothing, so his ill-equipped, understrength forces with the rules of engagement limited to use of force only in self-defense excluding intervention to protect civilians under attack, watched thousands of innocent unarmed civilians butchered in front of them in total violation of human rights and international humanitarian laws, which the United Nations unconditionally purport to uphold.

These thoughts flitted in and out of our minds as we drove along. But the forefront of our consciousness was the fate of the ten Belgian peacekeepers. Surrendered by Hutu *Interahamwe* militia, they called UN headquarters in Kigali for instructions. UN command, cognizant of the mission's inherent deficiencies such as limited strength, poor equipment and restrained rules of engagement that limited the capacity of UNAMIR forces to fight a sizable and well-equipped adversary, ordered them not to fight, indeed, to give up their arms.

The fates of these men meant so much to us because we were unarmed UN military observers sent there to obverse and report human rights violations and also assist in the settlement of returnees. Intimidation, snatching of mission vehicles, and robbing of mission personnel were common phenomena as well as the killing of prisoners and locals who furnished us with information on human rights violations. For example, in the presence of the locals, my team was arrested and held hostage for hours by the RPF/RPA who threatened to blast us together with our helicopter into pieces if we failed to give them accurate answers or made attempt to call UN headquarters to report our predicament. The locals complained and wondered why we could not protect ourselves and them so they refused to cooperate with us in reporting violations. We saw little local regard for the legal jurisdiction of men without guns and even those with small arms with rules of engagement limited to use of force only in self-defense but not to protect unarmed civilians under attack, despite our rather exalted international credentials. No doubt Rwanda supported ending the mission prematurely, stating that UNAMIR did not respond to its priority needs and UNAMIR left in March 1996.

Two years later I was in Sierra Leone as the Ghana Air Force Detachment Commander, as part of an Economic Commission of West African States Monitoring Group (ECOMOG). Sierra Leone had been pushed into civil war by Revolutionary United Front (RUF), an arm of Liberian Charles Taylor's "Greater Liberia"—a hidden military agenda to take over much of West Africa and make it a center for drug-running and diamond smuggling. Taylor had perfected the art of turning government into a criminal enterprise and unleashing armies of vicious, drug-besotted children to rape and slaughter for fun and terror. ECOMOG's enforcement action forced the rebels into the bush and we restored order and protected the populace from the onslaught of the RUF. As we took over previously RUF-held zones in the countryside afterward, we again saw unspeakable horrors . . . and hundreds of people, thousands throughout the country, whose hands or feet had been amputated by RUF "soldiers," apparently to cow the population into submission and make them withdraw their support for the legitimate government headed by President Tejan Kabba.

I was proud that I was a professional soldier who had helped to restore stability and peace to Sierra Leone. UNAMSIL, which was initially a traditional mission, and derided for its incapacitation to protect civilians and defending themselves instead, has been described as one of the United

Nations' success stories, surpassing the predictions of even some of the most enthusiastic optimists after it was later on handed a *robust mandate.* A *robust mission* had worked in Sierra Leone. I was immediately struck by the remarkable difference between the Rwandan mission and Sierra Leonean mission and by a comparison of their successes in protecting civilians. My concern, particularly, was the barbarity, amputations, and murders that were the price of traditional peacekeeping. Traditional peacekeeping, very much touted as a remarkable innovative conflict resolution mechanism in the cold war, failed to protect human rights in the post-cold war era violent civil wars exemplified by the Rwandese genocide. I wondered why we were successful in Sierra Leone but failed in Rwanda. It dawned on me that traditional peacekeeping, originally designed for interstate conflict, seemed an inappropriate tool for the post–Cold War civil wars involving irregular forces with total disregard for human rights and international humanitarian laws, whose *modus operandi* is to target civilians as a strategy in the pursuit of their objectives. But I wanted a more insightful knowledge about these mission types. Finding out the whys became my passion. I decided to earn a master of laws degree in international human rights specializing among others in international law of armed conflict and peacekeeping laws and also earned a doctorate in political science (international relations, American government, public policy and conflict resolution) to explore and contribute to the field of conflict resolution and human rights, particularly, civilian protection.

Acknowledgments

To achieve this career height required the tremendous support of a number of incredible individuals. I thank my family, especially sister Joyce and brother Patrick, and all my friends and loved ones, especially, Theresa Fountain-Coleman, and DaCosta Adu-Poku, for providing the necessary emotional and physical support throughout this academic journey.

I would like to give a special thanks to Dr. Frederic Pearson, my committee chair, advisor, mentor, and employer under whose supervision and persistence to excellence have helped me excel in academia. I thank the members of my dissertation committee; Dr. Daniel Geller, Dr. Timothy Carter and Dr. Osaka Likaka and my friend Susumu Suzuki for their insightful suggestions, support and encouragement. I am very grateful to Professor Marie Olson Lounsbery for reviewing the manuscript and endorsing the book. I am also very grateful to Professor Sir Nigel Rodley for your support during my Master of Laws (LLM International Human rights law) program at Essex University, UK. I also want to thank the faculty, staff and my colleagues of the department of History, Politics and Justice, Ohio Northern University, USA, for their generous support and encouragement. The work would not be nearly as strong without the thoughtful edits from Melanie Caw Woods, Tom Webb and Paula Scott, of the communication skills center, Ohio Northern University.

I wish I could detail how each person in my life made my dreams a reality, but the following individuals have my utmost gratitude and appreciation:

Philip Adu-Adjei, Tawia Bio, Francis Arthur, Peter Arthur, Kofi Donkor, Kofi Gyabeng, Kofi Gyenin, Vinep Kankam da costa, Group Captain Frank Hansen, Group Captain Kojo Ankah, Dr. Mary Duncan, Isaac Owusu Acheampong, Justice, Esther Yakubu, Felicia Kyeremateng, and Michelle Hamilton.

I will always be grateful for their love, patience, and support.

1

Introduction

Six questions drive this study: What accounts for United Nations (UN) peacekeeping forces' inability to protect civilians from human rights violations, especially intentional killings? Is the traditional peacekeeping operational doctrine, initially designed for interstate conflict, appropriate for the new intrastate conflict environment? Do mission characteristics of UN peacekeeping mandates and operations matter in reducing the ongoing intentional civilian killing? When deployed in civil wars, do UN peacekeeping missions work better than other mission types in staving off civilian killings? To what extent does robust peacekeeping affect the number of intentional civilian killings? Is there such a discernable outcome with respect to other stated peacekeeping mission goals on a case-by-case basis? Examining these questions will illuminate the defects within the traditional UN peacekeeping operational paradigm and the appropriateness of a new operational doctrine to boost UN peacekeeping forces' capability to protect civilians under threat in civil wars. The locus of the problem is that the United Nations seems to apply a traditional peacekeeping operational paradigm suitable for interstate wars to a post–Cold War environment of intrastate wars, which involve both state and nonstate actors whose modus operandi is to attack civilians to achieve their strategic political goals. Traumatic UN failures in Rwanda and Bosnia necessitated the reassessment of the fundamental principles of traditional UN peacekeeping and the suggestion of robust peacekeeping[1] This book examines the emergent conviction that traditional UN peacekeeping has largely failed and that robust peacekeeping, based on a strong and forceful peacekeeping force, is a better option for reducing human rights violations, especially civilian killings, in contemporary, volatile civil wars.[2] A few scholars and policy makers have discussed robust peacekeeping, but none has subjected it to empirical analysis as this book does. This book systematically analyzes the varied factors underlying the success and failure of missions, particularly in terms of human rights observance, using both statistical and case study models.

However, some officials and scholars who are of the opinion that robust peacekeeping would rather lead to more civilian killings have contested this approach.[3] Previous studies in the field focused on long-term effects after the termination of a mission and determined mission success using various dependent variables such as peace agreements and conflict termination.[4] Most of these studies did not focus on the relation between these variables and the immediate success in protecting civilians from atrocities. The missing link is that there has been limited study on how the characteristics of a UN peacekeeping mission affect ongoing civilian killings. This book sets out to provide that missing link. The book uses "success" in the sense of a reduction in human rights violations—specifically in the deliberate killing of civilians—while the peacekeeping mission is deployed. It focuses on immediate effects complements the works of others who emphasize long-term effects after the termination of a mission. This focus on immediate effects is justified on the basis that long-term goals, such as addressing the underlying causes of the conflicts and war, can be more successfully pursued in an atmosphere of immediate peace secured through peacekeeping missions. Purposeful and meaningful peacemaking and peace building cannot take place without an immediate and initial secure and enabling atmosphere. This book is a pioneering piece of work and the results of the study on which the book is based are astonishing. I believe this research conveys new and important information and sheds light on the operational capabilities of UN robust missions to protect vulnerable civilians from deliberate killings by spoilers in violent civil wars. I also believe the book will be of great interest to UN peacekeeping policy makers and practitioners and will provoke debate among multiple constituencies. Suggestions by UN policy makers and peace practitioners to adopt robust UN peacekeeping signify that the research is topical and of interest not only to the academic community but also to the informed public.

Methodology

The book examines the causes of variation in international peacekeepers' ability to prevent human rights violations, as well as the effect of robust peacekeeping on reducing human rights violations, specifically deliberate civilian killings, in areas of deployment. In view of the methodological divide in scholarly works between case study analysis and Large *N* quantitative aggregate data analysis, I adopt a mixed method called the nested analysis. The method combines case study analysis with statistical analysis, thus offering the advantages of each approach and also making up for the shortcomings of each. The methods are logit regression and negative binomial statistical models and focused structured case study comparison.

Specifically, the book examines the contributions of key structural variables, including mission type, major power participation, weapons type, rules of engagement, mission strength, and resources (funds), controlling for troop composition, regime type, peace agreements, borders, conflict type, and ethnic similarity between peacekeeping forces and host nations, using both qualitative and quantitative explanatory (negative binomial and logit regression) models. The basic argument is that various UN peacekeeping mission characteristics could have different impacts on civilian killings during civil conflicts. The book examines different types of UN peacekeeping missions—observer, traditional, multidimensional, and robust—in civil conflicts from 1956 to 2006. Included in the study are missions that recorded deliberate civilian killing of 25 persons or more per year. The unit of analysis is UN mission year.

Case Study

While a statistical analysis may give us a fair idea of how robust peacekeeping works, it may not capture all the factors impacting civilian killings; a detailed study of individual cases would help accomplish this. A closer look at individual cases therefore serves to inform the author as well as the reader of other potential significant factors impacting civilian killings. Cases of both traditional and robust peacekeeping missions are comparatively examined to establish other significant factors, such as lack of interest from major powers, that impact civilian killings but not captured in the statistical models. The book examines key representative cases of robust peacekeeping in Africa, including the Democratic Republic of Congo (MONUC) and Sierra Leone (UNAMSIL), and compares them with some cases of traditional peacekeeping, such as Rwanda (UNAMIR) and Angola (UNAVEMs), in the context of the formidable barrier model. These representative cases were chosen from the same region in order to minimize cultural and other biases, and they were chosen because the excesses in terms of human rights violations in these conflicts attracted the attention of both global scholars and policy makers. They also share other similarities.

Formidable Barrier Model of Robust Peacekeeping Success

The book formulates in Chapter 3 a new operational paradigm of UN peacekeeping mission success called the *formidable barrier* model of UN peacekeeping mission success. It develops a simple game-theoretic model to explain the relevance of robust peacekeeping as a guide to UN peacekeeping practitioners and state foreign policy makers. Borrowing from cost-benefit theories, the *formidable barrier* model posits that war, as well as deliberate civilian killings, is a rational process involving careful, deliberate calculations by rational actors. A group's (spoilers) probability of success or failure and its perception of the benefits and costs of continued fighting are a function of its capabilities relative to that of the defender (UN mission forces). Spoilers will continue to fight and kill civilians when they believe they are stronger than the mission force, but they do not do so when they are weaker.

The book argues that robust peacekeeping, due to its sturdy characteristics, has the capacity to constitute a formidable barrier and encourage spoilers to choose negotiations instead of war and killing of civilians. Robust peacekeeping is an appropriate peace operational paradigm, the book argues, because studies in the conflict management literature point out that an intense conflict with attendant human rights violations undermines the prospects of mediation success.[5] Until violence is stopped or at least managed, it is unlikely that any attempts to resolve competing interests underlying the conflict will succeed. Thus, long-term goals such as addressing the underlying causes of the conflicts and war can be more successfully pursued in an atmosphere of immediate, secure peace achieved through UN robust peacekeeping missions. Robust peacekeeping can affect the combatants' calculation by imposing military constraints while offering peaceful opportunities. Following the first landmark Security Council resolutions 1265 (1999) and 1296 (2000), which addressed the issue of civilian protection in armed conflicts,[6] the Security Council adopted resolution 1313, which authorized The United Nations Mission in Sierra Leone (UNAMSIL) "to deter and decisively counter the threat of RUF attack by responding robustly to any hostile action or threat of imminent and direct use of force."[7] UNAMSIL forces became the first peacekeepers to protect the Sierra Leonean populace from atrocities by the marauding Revolutionary United Front (RUF) rebels. Since then most UN peacekeeping missions have had robust mandates to protect civilians.[8]

The book has eight chapters. Chapter 1 introduces the book by providing a brief overview of the book's purpose. Chapter 2 examines the concept of human rights and UN traditional peacekeeping missions' ability to protect civilians from human rights violations by spoilers when deployed in civil wars. Chapter 3 discusses the literature on the subject, particularly on the evolving ideas of "robust" peacekeeping and also presents the conception of a model, the *formidable barrier* model, of peacekeeping success derived from the rational choice theory of conflict termination. This model posits that the success of peacekeeping missions in staving off deliberate civilian killings is a function of its deterrent capacity. It is argued that robust peacekeeping, by designation, has the ability to provide the necessary barriers or impediment to intentional civilian killings by spoilers by imposing some military constraints on them. From this model, I articulate a set of hypotheses and test them in Chapter 5 using UN peacekeeping data in civil conflicts from 1956 to 2006. Chapter 4 details the research models, methodology, data sources, and measurement of variables that include intentional civilian killings (dependent variable), mission type, mission size, weapon type, resources, rules of engagement, and major power participation(independent variables) controlling for other variables. In view of the methodological divide in scholarly works between case study and Large *N* quantitative aggregate data analysis, I adopt the mixed method called the nested analysis.[9] The method combines case-study analysis with statistical analysis, thus offering the advantages of each approach and also redressing the shortcomings of each. Chapter 5 provides an aggregate statistical analysis of the effectiveness of robust peacekeeping and civilian protection in civil wars. The empirical results indicate that the core variable "robust peacekeeping" has impact on civilian killings by lowering them. The strength of mission size associated with lower numbers of civilian killings seems to be the key factor. Major power participation, peacekeeper diversity, and affinity with the host state, along with identity conflicts and at least a proto-democratic status of the host state, appear to be harbingers of potentially higher deliberate civilian killing total.

Chapter 6 tests the *formidable barrier* model of robust peacekeeping, examining key representative cases of robust peacekeeping in Africa that include Democratic Republic of Congo (MONUC) and Sierra Leone (UNAMSIL) compared with some cases of traditional peacekeeping in Rwanda (UNAMIR) and Angola (UNAVEMs) in the context of the *formidable barrier* model. It was observed that UNAMSIL posed a formidable barrier whereas MONUC presented mixed results due to some factors not captured in my statistical model in Chapter 5. UNAVEM and UNAMIR, the key representative cases of traditional peacekeeping in Africa, did not constitute formidable barriers to the killing of civilians by the spoilers. The case studies captured relevant factors such as late deployment, mission force partiality, crimes by mission forces, porous border, poor topography and country size, excluded in my quantitative analyses, have adverse impact on the operational capacity of some missions in reducing intentional civilian killings. Using Darfur as a case study, Chapter 7 examines the politics within the United Nations, especially the Security Council, and the UN members' lack of political will to authorize the deployment of a robust peacekeeping mission. I argue that neorealist considerations of national interest hastened UN members' failure to support the deployment of a more credible robust peacekeeping mission in Darfur. Chapter 8 looks at the way forward in respect of UN robust peacekeeping and civilian protection in civil wars. I make policy recommendations such as the formation of a Peacekeeping Council comprising experienced troop contributing countries and practitioners to coordinate resources, improve training and supervision, analyze past and ongoing peacekeeping missions, and submit report as well recommendations to the UN Security Council. Other recommendations are resource provision, early deployment, adoption of standardized operation doctrine, training, rules of engagement and major power participation in UN peacekeeping missions, implementation the concept of responsibility to protect, and American leadership.

2

Peacekeeping and Human Rights Protection

The post–Cold War world has witnessed devastative armed conflicts replete with systematic violence and egregious human rights violations against civilians. Civilians today are the vast majority of casualties in armed conflicts and are increasingly targeted by combatants. Acts of violence against them, especially vulnerable women, children, and the elderly, include systematic killings, planned ethnic cleansing, mass rapes, starvation, disappearances, amputation of limbs, and torture. The nature and effects of human rights violations committed in civil conflicts remain disturbing and the United Nations, particularly the UN peacekeeping missions deployed in violent civil wars, has increasingly been expected to prevent and/or halt such crimes. Past UN missions designed to fulfill traditional peacekeeping roles in interstate wars were woefully underequipped, and both politically and operationally unprepared to take on the very challenging tasks of protecting vulnerable civilians in the changed post–Cold War violent civil conflict environment. The United Nations' traumatic experiences in the 1990s, when it failed to protect civilians from genocide in Rwanda, crimes against humanity in the former Yugoslavia, and in Darfur and the Democratic Republic of Congo, necessitated a reassessment of the fundamental principles and capabilities of UN peacekeeping operations. These failures demonstrate the urgent need for reforms in United Nations' peace operational paradigm to enable UN peacekeepers to play a much more active role in the protection of civilians when deployed in violent civil conflicts. Expectations of UN peacekeeping operations have steadily heightened in recent times, leading to the framing and adoption of UN Charter Chapter VII robust mandates, which routinely include authorization for peacekeepers to take measures to protect civilians under imminent threat of violence.

The protection and advancement of human rights is a fundamental objective of the United Nations in its avowed commitment "to save future generations from the scourge of war."[1] Member States affirmed their "faith in fundamental human rights, in the dignity and worth of the human person, in the equal rights of men and women" [2] with the ratification of the UN Charter. The United Nations has accordingly established numerous international human rights and humanitarian legal instruments to achieve this objective but the Cold War era historic superpower ideological antagonism militated against the United Nations' capacity, particularly the Security Council's, to effectively respond to human rights violations and threats to international peace and security. The end of the superpower confrontation was perceived as a new world order based on the ideals of international solidarity and respect for human rights,[3] which have figured prominently in the debates and activities of the United Nations.

UN peacekeeping is one of the most laudable conflict management mechanisms the United Nations has designed, originally for interstate conflict, to protect human rights during the Cold War when the paralysis of the Security Council hindered its ability to respond to global security threats. Post–Cold War era minimal superpower antagonism within the Security Council revived optimism about the United Nations' capacity to respond to atrocities especially against civilians in civil wars. It reinvigorated the Security Council's activism in peacekeeping missions leading to the deployment of robust missions today. However, UN record in promoting fundamental human rights has been abysmal and riddled with failure. The much vaunted optimism has been crippled by egregious conscience-shocking crimes against civilians reminiscent of the dark ages, even in the presence of UN peacekeeping forces deployed to operate in such civil wars as the Rwandese and Bosnian genocides. This has induced doubts in the consciousness of the global community on the viability of United Nations' contemporary peace operational doctrine in protecting human rights. It has consequently ignited scholarly, policy makers' discussions and the demand for a more competent operational paradigm to stem human rights violations in civil wars. This book evaluates the effectiveness of UN traditional peacekeeping operational doctrine and the emergent robust doctrine in protecting the human rights of civilians when UN mission forces are deployed in civil wars.

This book is based on the emergent conviction that traditional peacekeeping has largely failed and that robust peacekeeping, based on a strong and forceful peacekeeping force, is a better option for reducing human rights violations, especially civilian killings, in contemporary, volatile civil wars.[4] The central argument of this book is that robust peacekeeping works better than traditional peacekeeping in protecting civilians from killings. Borrowing from the rational choice theory of conflict termination, I develop the *formidable barrier* model of UN robust peacekeeping mission success, arguing that UN peacekeeping mission success would be attained if the mission has the capability to alter the calculations of conflict parties to choose peaceful negotiations instead of continued fighting. The objective of the book is to evaluate the ameliorative effect of UN peacekeeping missions on intentional civilian killings during civil wars through examining different types of UN peacekeeping missions—observer, traditional, multidimensional, and robust—in civil wars from 1956 to 2006 with the view to making certain policy recommendations. It is my hope that the answers will shed light on the operational capabilities of the UN missions in respect of human rights protection. Since human rights are at the heart of the United Nations and correlates with international peace and security, I discuss in the next two sections the concept of human rights and obstacles to its global implementation.

Human Rights

Human rights are those inherent, inalienable, and fundamental rights that all people are and should be entitled to irrespective of geographical and cultural diversity, violation of which is a legal wrong. [5] They are not privileges, hence legally irrevocable, because they are inherent in all humans[6] and reflect a society's principal "standard threats" to human dignity.[7] Human rights discourse articulates equality, justice, tolerance, mutual respect, respect for human dignity, and decent humane treatment of all persons of the world and thus adopts universalist and egalitarian postures reflecting shared norms of human moralities. They are grounded in the philosophical thoughts of some eminent philosophers opposed to absolutism and dogmatism, particularly the seventeenth-century English philosopher John Locke—arguably the most important natural law theorist of modern times. According to Locke, certain rights self-evidently pertain to individuals as human beings because they existed in the "state of nature" before they entered civil society. Prominent among these human rights are the rights to life, liberty (freedom from arbitrary rule), and property. Humankind upon entering civil society pursuant to a "social contract" surrendered to the state only the right to enforce these natural rights. A state's failure to honor its commitment to safeguard the people's reserved natural rights could give rise to popular revolution.[8] Following Locke, other liberal philosophers conceptualized universal principles of the "rights of man" that engendered struggles against political absolutism in the late eighteenth and nineteenth centuries.[9] America's 1776 Declaration of Independence, which proclaimed "We hold these truths to be self-evident, that they are endowed by their Creator with certain inalienable rights, that among these are life, Liberty and Pursuit of Happiness,"[10] was informed by these liberal philosophical concepts. France's 1789 Declaration of the Rights of Man and of the Citizens, which proclaimed "men are born and remain free and equal in rights,"[11] was similarly influenced by this liberal egalitarian intellectualism.

Locke's doctrinal conceptualization of human rights as natural rights was opposed by some liberal philosophers and ideologues such as Edmund Burke, David Hume, and Jeremy Bentham[12] who thought that such universalist and egalitarian posture and associated terminologies such as "universal," "inalienable," "irrevocable," and "eternal" connoted religious orthodoxy and absolutism.[13] Burke and Hume feared that popular affirmation of natural rights would promote social upheavals[14] while Bentham was concerned that proclamations of natural rights would substitute for effective legislation.[15] Burke condemned the drafters of France's Declaration of the Rights of Man and of the Citizens as "monstrous fiction" of human equality that inspires "false ideas and vain expectations in men . . . ."[16] Bentham, disputing the structural linkage of human rights and natural rights, wrote "Rights is the child of law; from real law come real rights; but from imaginary laws, (law of nature) come imaginary rights . . . natural rights are simply nonsense . . . upon stilts."[17] Relativists such as German jurists Friedrich Karl von Savigny and England's Sir Henry Maine also conceptualized human rights as derivative of the culture and environment of particular communities.[18] With such powerful philosophical and political attack on the doctrine of human rights as natural rights, the hope for "rights of man" was dimmed by Word War I.

Concern for international law on the protection of human rights is a post-1945 phenomenon when Nazi atrocities ignited active global consciousness to protect human rights. Before that, the protection of individual rights was the sole prerogative of individual states. The only exceptions were the issues of slavery (nineteenth century), trade, union rights, and certain minority rights.[19]

It must be admitted that the League of Nations, which had been unable to prevent the barbarous violations, had not been totally indifferent to human rights. The Covenant of the League of Nations, its founding treaty, referred to "fair and humane conditions of labor for men, women and children" in Article 23(a), to "just treatment of the native inhabitants of territories under their control" in Article 23(b). The League of Nations, however, lacked a broader philosophical perspective of universalizing human rights and linking respect for human rights to global peace as articulated in the UN Charter.[20] It was in the post–World War II era that the United Nations Member States declared their determination "to reaffirm faith in fundamental human rights, in dignity and worth of the human person" by signing the UN Charter.[21] Events leading to World War II (for example, Hitler's human rights violations) focused the global community's attention on the guarantee of human rights. Human rights then subsequently became an issue of global and interstate concern with the advent of the UN Charter. We subsequently observed increasing acceptance of the universalist, egalitarian, and inalienable ideals of human rights that are to be recognized by all and need not be conferred by states because they are fundamental to human dignity. Human rights gained primacy in UN operational paradigm reflected in the UN Charter and in associated conventions such as the Universal Declarations of Human Rights.

The realization, protection, and promotion of human rights, through international cooperation, became the purpose of the United Nations proclaimed in Article 1 of the Charter.[22] UN members in Article 56 of the Charter pledged "to take joint and separate action in cooperation with the Organization" to achieve the purpose in Article 1 of the Charter.[23] Both the General Assembly and Economic and Social Council (ECOSOC) are charged to make recommendations to promote human rights in Articles 13 and 62 of the Charter. Of special significance is the Universal Declaration of Human Rights. Human rights have also been of special concern to nongovernmental organizations accredited to the United Nations. Despite controversy over the nature and interpretation of state sovereignty norms,[24] the international community has become increasingly aware of its global responsibilities in the field of human rights. Article 24(1) of the United Nations confers on the Security Council the primary responsibility for the maintenance of international peace and for that matter specific powers for the peaceful settlement of disputes under Chapter VI of the UN Charter and to take action in respect of threats to the peace and acts of aggression under Chapter VII of the Charter.[25]

Despite its seemingly universal acceptance reflected in the ratification of the UN Charter and Universal Declaration of Human Rights (UDHR) by UN members, human rights discourse has been dogged with controversies and debates over the content, legitimacy of some rights, and prioritization of some rights over others. Of particular significance is French jurist Karel Vasak's categorization of human rights into three generations, informed by the three themes of the French revolution: Liberty, Equality, and Fraternity.[26] The controversy over the prioritization of the three generations rests on the political divisions of the Cold War. First-generation human rights deal fundamentally with civil and political rights; the second generation covers economic, social, and cultural rights; and the third-generation rights are called solidarity rights. First-generation human rights are fundamentally civil and political in nature and essentially deal with individual citizen's liberty and political participation. This generation of human rights developed under a strong mistrust of government. Historically informed by Locke's natural law, and political philosophy of individual liberalism, first-generation human rights uphold the principles of individualism and governmental noninterference to protect the individual citizenry against the abuse and misuse of state political authority and to unrestrictedly exercise those rights. The state's relevant involvement is the creation of a serene enabling environment for citizens to exercise these rights. Some first-generation human rights, however, such as rights to security of a person, to a free and fair trial, and to free and fair election require governmental interference to guarantee them. First-generation human rights, based on the Anglo-American principles of liberty, are mostly "negative" rights—rights from certain things, usually freedoms from abuse such as torture or coercion racial discrimination.[27] Informed by Hobbesian–Lockean liberal individualism of human rights, first-generation human rights are ideologically prioritized and promoted by Western democracies. These rights are articulated in the US Bill of Rights; France's Declaration of the Rights of Man and of the Citizens in the eighteenth century; Articles 3–21 of the 1948 Universal Declaration of Human Rights.

Second-generation human rights are economic, social, and cultural rights in nature. Informed by Marxist and radical traditions of the nineteenth-century revolutions and welfare movements, these rights were developed in opposition to capitalism's devastation and exploitation of the working class.[28] In response to capitalism, the emergence of working class as a formidable political force and its subsequent political struggle led to fundamental changes in the conceptualization of conditions befitting a dignified life.[29] European states had dominated the international political system and influenced international norms and laws from the period when The Peace Treaty of Westphalia was signed in 1648 to the end of World War I. The industrial revolution in Europe gave them strategic advantage over non-European states that lived on the periphery of the international community without any major role in global politics, particularly the formulation of international norms and laws. International law and norms then was the product of Western civilization based on Eurocentricism, Christian ideology, and free market outlook. The emergence and relevance of Communist and Third World states in global politics at the end of the World War I and the collapse of colonialism, respectively, led to agitation for changes in international status quo and normative laws to include their concerns and interests. International law, they argued, were upshots of "bourgeois" and capitalism and therefore contrary to their interests.

Second-generation human rights were thus upshots of demands for normative structural changes in the international system. This generation of rights was generated, pioneered, and promoted by the socialist communists and Third World states to cater to their interests and normative values, and counterbalance Eurocentric first-generation human rights. They are based on the principles of distributive social justice and public obligation focusing on equal conditions and treatment of citizens by the state. Second-generation human rights conceptualizes human rights in positive terms (rights to) requiring state intervention to institutionalize appropriate measures to guarantee equal opportunities to all the citizenry and provide for its neediest citizenry.[30] These rights are articulated in Articles 22–27 of the Universal Declaration of Human Rights. The International Covenant on Economic, Social and Cultural Rights includes the right to social security; the rights to work and to protection against unemployment; the right to rest and leisure, including periodic holidays with pay; the right to a standard of living adequate to health and well-being of self and family; the right to education; and the right to the protection of one's scientific, literacy, and artistic production. Despite liberal democracies' prioritization of first-generation human rights, they have implemented measured policies to promote second-generation rights. US president Franklin D. Roosevelt proposed a Second Bill of Rights protecting second-generation human rights during his State of the Union Address on January 11, 1944.[31] European Union's European Social Charter is a legally binding declaration guaranteeing comprehensive sets of second-generation human rights.[32]

Third-generation human rights are group and solidarity rights that promote the collective interests of the Third World and rights of vulnerable groups in society such as children, women, the disabled, and ethnic minorities. These rights are articulated in Article 28 of the Universal Declaration of Human Rights that proclaims that "everyone is entitled to a social and international order in which the rights set forth in this Declaration can be fully realized."[33] Other progressive documents of international law such as the 1972 Stockholm Declaration of the United Nations Conference on the Human Environment,[34] and the 1992 Rio Declaration on Environment and Development also articulate these rights.[35] These rights include the right to socio-politico-economic self-determination; the right to economic and social development; the right to natural resources; and the right to participate in and benefit from the "common heritage of mankind" (shared Earth-space resources: scientific, technical, and other information and progress). Third-generation human rights were fuelled by third world nationalism and associated demands for global redistribution of power, wealth, and other important values relevant to them. However, most of these third-generation rights, as also some second-generation human rights, remain more aspirational than justifiable because their protection hinges on resource availability. Yet, some countries and international organizations have institutionalized certain measures to safeguard third-generation rights. The Hungarian Parliamentary Commissioner for Future Generations[36] and Finland's Parliamentary Committee for the Future[37] are constitutional mechanisms safeguarding third-generation rights. The High Commissioner on National Minorities of the Organization for Security and Co-operation in Europe safeguards the rights of minorities.[38] The United Nations undoubtedly faces a herculean task in the protection of human rights due to daunting obstacles undermining its capacity to achieve its objective.

Obstacles to Human Rights Protection

Intense philosophical debates and criticisms of the concept of human rights have precipitated formidable obstacles to international efforts to promote and protect human rights globally.

Prioritization of Generations of Human Rights

The first obstacle is the contestation among the UN Member States over the legitimacy and prioritization of the generations of human rights. The disagreement over generations of human rights rests on the political divisions of the Cold War fuelled by ideological biases of Western democratic individualism and communist collectivism. The sequel is that these rights are neither accepted nor given equal urgency. Western democracies insist on civil and political rights first, whereas communists and Third World states defer it until the well-being of their people is assured. Proponents of first-generation human rights, informed by Hobbesian–Lockean liberal individualism of human right, prioritize them over the second-generation rights because of the ease and flexibility of implementing it independently of economic prosperity or resource availability and it is void of governmental intervention. These rights, they argue, can easily be provided if the state decides to. First-generation human rights are said to be justiciable because their provision and breach of them can be judged in court of law, and therefore, constitute real "legal" rights. They are precise since their provision is easy to judge and measure. Many Western conservatives and philosophers, such as Maurice Cranston, disparage most of the economic and social rights. Cranston argues that civil and political rights are "universal, paramount, categorical moral rights"[39] and economic and social rights "belong to a different logical category"[40] and therefore are not truly human rights.[41] He states that such rights "do not make sense" and suggests that claims of such rights probably are not even intelligent.[42] According to Cranston, political and civil rights "generally . . . can be secured by fairly simple legislation"[43] whereas "economic and social rights can rarely, if ever, be secured by legislation alone."[44] It is also cheaper to implement and realize civil and political rights whereas the guarantee of economic and social rights is costly, and even may be impossible, to implement in some poor countries due to resource shortages.[45] Social and economic rights are viewed as utopian aspirations[46] and scarcity connotes that they are not really rights at all. Second- and third-generation human rights are said to be vague because they cannot be quantitatively measured and it is difficult to judge whether they are adequately provided or not. They are non-justiciable because either provision or the breach of them cannot be judged in a court of law since they are resource intensive. They are therefore programmatic so as to be realized gradually and therefore do not meet the threshold of human rights. Conclusively then, second- and third-generation human rights are rather aspirations or goals than real "legal" rights since countries are not equally resource endowed and poor states cannot be legally held accountable for failing to provide employment, food, and houses to its poor citizens. This differs from first-generation human rights whose exercise does not hinge on resource availability but is simply based on governmental noninterference and the provision of enabling environment for the exercise of those rights. States therefore lack any legitimate excuse for failing to guarantee first-generation human rights and can therefore be held accountable for such deliberate failures. According to liberal political philosophers and policy makers, non-Western governments' excuse of prioritizing second-generation human rights over first-generation human rights for the sake of solidifying national efforts for national economic development is a screen to rationalize authoritarian governance.[47] Proponents of second generation in rebuttal argue that Western capitalist democracies that prioritize first-generation human rights over second-generation human rights are inattentive to the material needs and development of the human capacity of their citizens. Western democracies' fixation on liberty and individualism embodied in first-generation human rights are stratagems to rationalize the exploitation and other abuses of capitalism[48] that promote social injustices and exploitation of the disadvantaged citizenry.

International human rights instruments, however, do not legitimize and prioritize some human rights over others except *jus cogens*, which are non-derogable and more fundamental than others such as freedom from torture. The 1993 Vienna World Conference on Human Rights conclusively resolved the debate by emphasizing the principles of universality, indivisibility, and interdependence of human rights when it declared that

> All human rights are universal, indivisible and interdependent and interrelated. The international community must treat human rights globally in fair and equal manner, on the same footing, and with the same emphasis. While the significance of national and regional particularities and various historical, cultural and religious backgrounds must be borne in mind, it is the duty of States, regardless of their political, economic and cultural systems, to promote and protect all human rights and fundamental freedoms[49]

The International Covenant on Civil and Political Rights (ICCPR) also emphasizes the principles of indivisibility and interdependence of human rights in its preamble, stating that

> in accordance with the Universal Declaration of Human Rights, the ideal of free human beings enjoying freedom from fear and want can only be achieved if conditions are created whereby everyone may enjoy his or her economic, social and cultural rights, as well as his or her civil and political rights.[50]

The Declaration on the Right to Development adopted by the General Assembly of the UN on December 4, 1986, also underscores interdependence and indivisibility of human rights. The basic ideas in this Declaration include the acknowledgment that all human rights are indivisible and interdependent; and the realization that the failure to observe civil and political rights, as well as economic, social, and cultural rights constitutes obstacles to development.[51] The principles of indivisibility and interdependence imply that both generations of human rights are equally important and their promotion and protection would guarantee decent and dignified lives for humankind. The reality is, some of these rights are interdependent and their enjoyments are structurally linked. Some first-generation human rights require second-generation human rights to thrive and vice versa. For example, the second-generation right to education may be essential to first-generation right of freedom of speech, religion, and political participation. Economic and social rights to food, health care, and housing are essential to civil and political right to life. Prioritization and legitimization of certain rights and neglect of others would therefore undermine the sharing of all values in an interdependent globalized civil society and such prioritization and legitimation do not reflect global reality.[52]

In accordance with these realizations, most Western democracies, despite prioritizing first-generation rights, uphold second-generation human rights. US Secretary of State Cyrus R. Vance in 1977 acknowledged the relevance of second-generation human rights when he defined human rights to include "the right to be free from governmental violation of integrity of the person ... The right to the fulfillment of such vital needs as food, shelter, health care, and education ... the right to enjoy civil and political liberties."[53] The United States has accordingly institutionalized welfare programs such as food stamps, social security, and Medicare and Medicaid for its citizens. Similarly, communal societies such as China tolerate and even promote certain individual rights such as the right to property and right to life. Despite the near-unanimity on the interdependence and indivisibility of human rights, Member States' philosophical and ideological differences and preferences continue to hamper United Nations' activism to promote all human rights. Second-generation human rights still enjoy less observance in most democracies as their breaches do not invoke the same response as when first-generation human rights are violated. On the hand, communal societies such as China and developing states fixated on economic development prioritize second-generation human rights over first-generation human rights, which they accuse of being divisive and would derail impoverished states' collective efforts toward development.

Universalism–Relativism Debate

The universalism–relativism debate in human rights discourse undermines universal protection of human rights. Universalists, sympathetic to Western liberal traditions, claim that all human beings, irrespective of cultural or regional differences, are entitled to human rights. Human rights, thus, transcend all cultures and are the same everywhere and all governments have obligations to implement them. International human rights discourse and instruments such as the 1948 UDHR, the ICCPR, and the International Covenant on Economic, Social and Cultural Rights (ICESCR) have adopted universalist postures. They envision that equal human beings are endowed with inalienable rights that entitle them to equal concern and respect from states.[54] The ICCPR "recognize that these rights derive from the inherent dignity of the human person"[55] and the UDHR emphasizes the inalienable and inviolable rights of all members of the human family and the international community's obligation to protect them.[56] The 1968 Proclamation of Tehran,[57] the Bangkok NGO Declaration on Human Rights,[58] the 1993 Vienna Declaration,[59] and the ratification of UDHR, ICCPR, and ICESCR by over 120 UN Member States from all regions of the world seem to support and affirm that universal human rights standards are deeply rooted in many cultures of the world.[60] Universalists denounce relativism as an alibi to justify violations of human rights by the ruling class that benefit from maintaining the status quo.[61]

The universalist line of human rights discourse has been criticized and resisted by relativists of different cultures and ideologues who argue that human rights are relative to cultures. According to Relativists, issues of morality that underlie human rights are derivatives of culture, hence the notions of human rights differ in a pluralistic international community with diverse cultures.[62] The scope of human rights for individuals is therefore determined by the cultural particularities of a given society.[63] Contemporary human rights arguably are the exclusive heritage of the Western liberal political tradition[64] that first emerged in the West in response to threats to human indignities.[65] The West suffered historical experiences of injustices and human indignities that include the Atlantic slave trade, savage colonialism, religious persecution, virulent racism, absolute monarchy, predatory capitalism, global warfare of almost unthinkable destructiveness, fascism, communist totalitarianism, and a host of other evils and social ills.[66] The West consequentially developed international human rights norms to confront these human indignities giving contemporary human rights discourse a Western twist. This has been fostered by the West's global military, political, economic, and social dominance and the absence of a formidable alternative, non-Western human rights discourse.[67]

Focusing primarily on the historic origins and philosophical underpinnings of UDHR and ICCPR, the two fundamental human rights instruments, Third World states argue that their concerns and interests were ignored because they were colonized by the West before human rights discourse evolved. Third World states, therefore, have difficulties accepting European standards of morality as the standards for all humanity and rejected the imposition of Eurocentric standards on other dissimilar cultures as not justifiable. Universalizing the Eurocentric bourgeois-liberal construct of human rights reflects Western arrogance or cultural imperialism,[68] which must be resisted. The seemingly voluntary ratification of international human rights instruments by non-Western states are therefore for political expediency. Non-Western authoritarian regimes, that are unaccountable to their citizenry, lack the moral legitimacy to represent the interest of their people. Their politically expedient ratification of international human rights treaties, therefore, does not reflect the interest of the people. Many non-Western authoritarian states, such as Zaire under Mobutu Sese Seku, "ratify human rights instruments to blunt criticism, protect sovereignty, keep themselves in power and have access to resources."[69] African relativists oppose the universality of human rights and criticize the individualistic premise of first-generation human rights discourse as contradictory to Africa's communalistic social life. In Africa, the individual is not isolated from the community, rather he or she is an integral member of the community anchored by a spirit of solidarity.[70] An individual's rights, needs, sorrows, and duties are structurally woven into the life of the collectivity that does not tolerate runaway individuals.[71] Individuals' best maximize their rights, interests, and duties within the group. The survival of the group is therefore paramount to the individual. Individuals in Africa are not "atomistic units locked in a constant struggle against society for the redemption of their rights."[72] The concept of group-centered individuality in Africa delicately entwines rights and duties, and harmonizes the individual with society.[73] African states are, therefore, not antagonistic to individuals but rather best maximize individual needs. Therefore, overindulgence of the individual right at the expense of collective rights is not permissible in African culture.[74] African leaders' modernization objective that prioritizes socioeconomic development of their impoverished masses over politics also significantly explains their infatuation with the relativist argument. Western exploitative colonialism delayed Africa's development far behind the rest of the world. To bridge this developmental gap requires a united front reminiscent of the anti-colonial movement era with least tolerance for divisions. Democratic political systems that entail multiple parties and guarantee individual civil and political rights are seen as divisive and therefore impediments to be sacrificed for the sake of the collective good.[75] A dictatorial one-party political system is considered the ultimate choice, and a multiparty system is demonized as counterproductive to development. Multiparty systems inherently foster divisions along ethnic or tribal fault lines, which poor African states do not have the luxury to tolerate at the very time that national unity is needed for national development. The sequel is draconian measures that violate the African people's first-generation human rights. These measures are not viewed as violations but they are rather seen as requisites to foster national unity for socioeconomic development of poor and ethnically diverse societies.[76]

The "Asian Values"–universalism debate also contributes significantly to the limited applicability of human rights in Asia. "Asian values" criticize Western-oriented contemporary human rights regimes committed to universalizing human rights and democracy as an aversion to the global reality of diversity and plurality. Singapore's foreign minister to the 1993 Vienna World Conference on Human Rights, for example, warned that "universal recognition of the ideal of human rights can be harmful if universalism is used to deny or mask the reality of diversity."[77] Asia is said to have a unique set of values emanating from its institutions and political ideologies that provide the basis for Asia's different understanding of human rights and justify the "exceptional handling of rights by Asian governments."[78] The 1993 Bangkok Declaration on Human Rights fully endorsed the Asian values that are strongly opposed to universal human rights by emphasizing Asia's commitment to the principles of sovereignty, self-determination, and noninterference.[79] Asian values articulate Asian societies' predisposition toward collectivism and communitarianism over individualism; preference for social harmony and consensus as opposed to confrontation and dissent; preference for the welfare and collective well-being of the community over individual rights and loyalty and respect for authority and authoritarian regimes as opposed to liberal democracy.[80] Universal human rights according to the "Asian view," are individualistic by nature and hence destructive of Asia's social system. Increasing rates of violent crime, family breakdown, homelessness, and drug abuse are cited as examples of moral decadence that Western individualism breeds, particularly the United States.[81] Some Asian governments, particularly China, Indonesia, Malaysia, and Singapore, justify their authoritarian systems citing their relative economic success attributable partly to their commitment to communitarianism rather than individualism; consensus-building rather than the adversarialism associated with Western democracies. Western states' criticism of Asia's human rights practices therefore violates their sovereignty and self-determination. The denial of individualistic civil and political rights, such as freedom of expression and association, arbitrary political imprisonment, cruelty to detainees, and restrictions on democratic processes, are deemed appropriate for socioeconomic development of the Asian people exploited and underdeveloped by Western colonialism. There are, however, questions about the legitimacy of ascribing common values to the whole of Asia considering its religious, ethnic, political, and economic diversities.[82] Asia lacks cultural, economic, religious, and political homogeneity. Asia harbors all the major religions of the world: Christianity in the Philippines; Islam in Malaysia, Hinduism in India, and Buddhism in Thailand. Politically, Asia exhibits dictatorship in Myanmar, democracy in India, and the communist regime in China. According to Ghai, perceptions of human rights in Asia are reflective of social and class positions in society and the generalized Asian perspective on human rights in reality is the perspective of the privileged ruling elite, which seeks to justify authoritarianism and repression in order to perpetuate the status quo.[83] Asian political elites, similar to their African counterparts, manipulate cultural imagery to further economic development so as to retain power.[84] African leaders' developments before individual rights mantra, as well as Asian leaders' "Asian values" argument, are strategic manipulative ploys of the exploitative ruling elite to ward off external and internal criticisms and protect their socio-politico-economic dominance over the impoverished and deprived masses who clamor for state respect of their civil and political rights, as evident in the contemporary Arab Spring uprisings.

Sovereignty, Domestic Jurisdiction, and Self-determination

The United Nations' dual promotion of universal human rights and the principles of state sovereignty and associated principles of domestic jurisdiction and self-determination are counterproductive. The principles of state sovereignty, domestic jurisdiction, and self-determination that guarantee UN Member States the rights to shape their own socio-politico-economic destinies without external dictation fundamentally constitute formidable barriers to the universal applicability of human rights. While the preamble of the UN Charter reaffirms the faith in human rights and Article 1 proclaims international cooperation to promote human rights as one of the purposes of the United Nations,[85] the UN Charter clearly upholds the concepts of state sovereignty and the associated principles of self-determination and domestic jurisdiction. One of the thorniest issues confronting the United Nations is how to balance these concepts with human rights. Sovereignty entitles states to external noninterference in their domestic affairs. A sovereign is a "person who received the habitual obedience of the members of an independent political society who did not owe such obedience to any person."[86] State sovereignty therefore means state autonomy in issues within its territory without external influence. The states are insulated from legal scrutiny and therefore from abroad in matters within the states rooted in the notion of domestic jurisdiction. According to Roth, the international juridical sense of "sovereignty" connotes equality of legal status with an implied presumptive duty on the part of members "to respect the outcome of political processes internal to the others"[87] without external interference. The operating principle of international legality, therefore, should be understood to be "bounded pluralism" rather than "transcendent justice."[88] In other words, the international arena should be seen to accommodate diverse modes of political organizations to preserve states' territorial integrity and political independence. Universalizing human rights would, therefore, be problematic in view of the plurality of political and ethical values in the international arena.

The provision of domestic jurisdiction in Article 2(7) of the UN Charter states, "nothing contained in the present Charter shall authorize the United Nations to intervene in matters which are essentially within the domestic jurisdiction of any state . . . " except with its consent. This evidently prohibits the United Nations to act in such matters. Domestic jurisdiction insulates states from legal scrutiny of the United Nations on a broad category of issues within a state.[89] In this context, the concept of domestic jurisdiction seems to obstruct the universal protection of human rights. Domestic jurisdiction thus operates as a set of legal limitations on the establishment and enforcement of international human rights norms without state consent. In Darfur's case the objection of the Sudanese government to the deployment of UN forces and the United Nations' continuous negotiations with the Sudanese government for consent for deployment seem to affirm Sudan's jurisdiction over the Darfur crisis despite the UN declaration that the Darfur crisis posed a threat to global peace and security. The idea of domestic jurisdiction against infringement by an international organization is implicit in international agreements where the agency can only perform functions to which states consent. The inclusion of reservations to treaties underscores the importance of domestic jurisdiction to states. At the initial acceptance of international jurisdiction, states include escape clauses that enable them to avoid international interferences in issues deemed domestic. The United States, considered the citadel of human rights, made certain reservations to the ICCPR with respect to hate speech, double jeopardy, and the death penalty, with the justification that the mentioned provisions were in conflict with its domestic law.[90]

The principle of self-determination guarantees nations the right to freely choose their sovereignty and international political status without external compulsion or external interference. Essentially, the right to self-determination is the right of a people to determine its own destiny. The principle of self-determination is embodied in Article 1(2) of the UN Charter that states that a purpose of the United Nations is "to develop friendly relations among nations based on respect for the principle of equal rights and self-determination of peoples, and to take other appropriate measures to strengthen universal peace."[91] It is recognized as a right of all peoples in the first article common to the ICCPR and the ICESCR, which both entered into force in 1976. Paragraph 1 of this Article provides: "All peoples have the right to self-determination. By virtue of that right they freely determine their political status and freely pursue their economic, social and cultural development."[92] In line with these analyses, some states argue that universalizing human rights unequivocally undermines states' fulfillment of the cherished principle of self-determination that fundamentally defines their very existence and survival.

Anarchic World and Major Powers' National Interests

Neorealism, a theory of international relations, posits that anarchy prevails in the international system without established international authoritative institutions to make and enforce international laws including human rights. Since no international organization is ultimately capable of enforcing order, every state indulges in self-help to provide for its own security by pursuing power, which realists argue is the ultimate interest of states.[93] The United Nations, the closest to a world government, mandated to make and enforce international human rights laws, has been incapacitated by the prioritization of individual Member States' parochial national interests over their collective interests. Article 39 under Chapter VII of the UN Charter empowers the Security Council to "determine any existence of any threat to the peace, breach of the peace, or act of aggression" that includes human rights violations and will take "measures . . . to maintain or restore international peace and security."[94] However, the Security Council's capacity to respond to global atrocities is undermined by the strategic rivalry fuelled by neorealist and ideological considerations of its five permanent members. The Cold War ideological contest between the West and the East paralyzed the Security Council's response to human rights violations. For the Security Council to function effectively, the concurrence of the so-called "Big Five" is always required. As the Cold War confrontation increased, it was almost impossible to achieve concurrence among the major powers. The United Nations could not act and the Security Council's responsibility to maintain international peace and security was marginalized. The two major powers during the Cold War confrontation even sustained their brutish ideological clients who impudently violated the rights of their citizens without remorse. US Cold War strategic interest was the containment of the Soviet Union and communism to avert Soviet hegemony. The United States therefore sought more strategic allies and overlooked atrocities committed by its dedicated clients opposed to communism such as Mobutu Sese Seko. The United States, for example, sustained the brutish regime in Zaire that plundered the economy for three decades.

Neorealist strategic security, economic, and other interests undergird the Security Council members' interventional decisions to protect human rights, not the severity of a tragedy.[95] It is expected that the Security Council and major powers' interventional decisions would be informed by the severity of a human tragedy. But while the United States, for example, had supported robust interventions in Kosovo and East Timor with comparatively less severe violations, the Rwandese genocide did not garner such attentions. The United States was very instrumental in the signing of the Arusha Accord that prescribed a robust mission for Rwanda. However, following the Somalia debacle in which 15 American troops were killed by Somali factions, United States' interest in the Rwandan crisis waned and the subsequent promulgation of the PDD 25 handed Rwanda a poorly designed and ill-equipped traditional mission that could neither protect it nor stop the genocide. Neorealist-driven discourse suggests that American leaders, like leaders of all other states, must ensure the physical safety of the citizens of their own country from foreign attacks as well as other interests (economic, cultural) in an anarchical international world. The United States as a state actor in this world is a rational actor and commitments to engage in humanitarian disasters are premised on a complex calculation of perceived interests. Kaplan underscores the neorealist discourse underlying US interventions in human rights crises when he states that the United States should only intervene in conflicts where "moral, economic, and strategic interests intersect."[96] Former Senator Sam Nunn, an influential and well-informed analyst of American strategy, clearly articulated the philosophical underpinnings of America's selective responses to human rights violations elsewhere when he declares "our role is too important in areas of the world that are significant to United States' military interests, security interests, and economic interests to allow our military effectiveness to be dissipated in places where we have no economic and no security interests."[97] China's neorealist strategic interests in Sudan have clouded the Council's decision on Darfur killings. China has resisted declaring the Darfur killings a genocide and has thwarted efforts by some members of the Security Council to deploy a robust mission in Darfur due to its economic dealings with the Khartoum government. China obtains oil from Sudan and has arms trade deals with the Sudan government accused of international crimes against the Darfurians.

Major power interests and contestations undermined the designation of potent human rights instruments to effectively punish violations. For example, the UN Charter, typical of major constitutive instruments, is conspicuously general and vague in its human rights clauses. Parties to the Charter accept nothing more than promotional obligation of human rights as the United Nations' enforcement capacity has been weakened by a multiplicity of derailing factors. The constellation of all these factors has conspired to undermine protection of human rights.

Human Rights Abuses in Armed Conflicts

The four 1949 Geneva Conventions and their 1977 Additional Protocols prohibit attacks on noncombatant civilians and all persons and oblige conflicting parties to spare and protect them. Despite these provisions, innocent civilians have been targeted for strategic reasons by conflict parties during wars. UN Secretary-General Kofi Annan acknowledges that "it is increasingly true that the main aim . . . [of conflicts] . . . is the destruction not of armies but of civilians and entire ethnic groups."[98] It is estimated that approximately 3.3 million civilians have lost their lives in the Democratic Republic of Congo (DRC) war since 1998.[99] In today's civil wars, rebels strategically attack civilians to influence government response and improve their bargaining position in the war.[100] Attacks on civilians are therefore stratagems to boost rebels' relative bargaining power by changing the calculations of the government to their demands when military means are inadequate to compel the government to make concessions.[101] Civilians are simply bargaining chips. Conflict parties attack civilian supporters of their opponents if it may impose costs on their opponents and force them to make concessions. Human rights violations against civilians include "crimes against humanity," including genocide, torture, slavery, rape, enforced, extra judicial killing and disappearances, and "war crimes" that are serious violations of the laws of war (*jus in bello*).[102] Sexual violence aims to humiliate men and the community, portraying them as weak and incapable of protecting their women. Rapes are used to silence politically active women and are designed as a part of genocidal strategy, as in the Bosnian and Darfur conflicts.[103] Torture, either physical or psychological, is widely carried out to humiliate and annihilate the dignity of the victim.[104] Physical torture might include mutilations, beatings, and electric shocks to lips, gums, and genitals, while psychological torture includes water-boarding, sleep deprivation, and denial of food and water for long periods, and exposure to high-powered lights.[105] Abuses of children include detention, deploying children as soldiers, and using children as sexual slaves. The Lord's Resistance Army (LRA) commanders, for example, made captured children take part in ritualized killing of others very soon after their abduction.[106] Ironically, these violations are sometimes committed in the presence of UN peacekeeping forces deployed in the conflict zones with an implicit mandate to protect civilians, such as the United Nations Assisted Mission in Rwanda (UNAMIR) that failed to protect Rwandese during the 1994 Rwandan genocide because of restricted rules of engagement that permitted UNAMIR forces to use force only in self-defense. UN peacekeeping forces were mere bystanders when Hutu extremists murdered some 800,000 Tutsis and politically moderate Hutus.

Peacekeeping

Peacekeeping is one of the United Nations' most innovative interventional mechanisms to manage and resolve conflicts. Peacekeeping fulfills the purposes of the United Nations in its task of maintaining international peace and security. Cold War confrontation made it almost impossible to achieve concurrence among the major powers paralyzing the Security

Council's ability to deal with conflicts that threatened international peace and security. The United Nations was weakened in performing its primary role of maintaining peace and security and explored less demanding new techniques to avoid and contain conflicts.[107] Peacekeeping as a concept and as a function of the United Nations was inaugurated during the difficult days of the Cold War in places such as Suez, with the intent of keeping the warring sides apart and obviating the necessity for major power intervention.[108] The evolution of peacekeeping is the sequel of the gradual abandonment of the idea of the use of military force by the United Nations to counter aggression, especially during the Suez crisis, to prevent a war between the major powers.[109] Early architects and exponents of peacekeeping included Lester Pearson, Canada's UN ambassador and Under-Secretary General, and well-known mediator Ralph Bunche (Appendix table 2.1 presents a list of all intrastate UN peacekeeping missions for the period of 1956–2006). Lester Pearson, as a compromise, initiated peacekeeping as a concept during the Suez crisis when the world was on the brink of war. During this crisis in 1956, Lester Pearson proposed "a United Nations force large enough to keep those borders at peace while a political settlement is being worked out."[110] The concept of "peacekeeping" operations evolved slowly and in an ad hoc manner, for there is no specific provision for it in the UN Charter. Peacekeeping is unique and the former UN Secretary-General Dag Hammarskjold referred to it as belonging to "Chapter Six and a Half" since it goes beyond traditional Chapter VII pacific measures, but stops short of Chapter VII collective security actions.[111] The UN Charter does not contain specific provisions for peacekeeping [112]and there is no agreed definition.[113] Jean Marie Guehenno, former Under-Secretary General for the United Nations Department of Peacekeeping Operations, remarks: "there is still no definition on what it entails no criteria when operations are to be established and no guidelines on how to plan and deploy the forces."[114] The United Nations defines peacekeeping as a unique and dynamic instrument developed by the organization as a way to help countries torn by conflict to create conditions for lasting peace.[115] For the purpose of this study, I adopt International Peace Academy's broad definition of peacekeeping "as the prevention, containment, moderation and termination of hostilities between or within states through the medium of third party intervention, organized and directed internationally, using multinational military, police and civilian personnel to restore and maintain peace."[116]

During the first UN peacekeeping mission, United Nations Emergency Force (UNEF) was created to facilitate withdrawal of troops, prevent a recurrence of hostilities, and act as a buffer separating the belligerents during the 1956 Suez crisis. Since then, the UN peacekeeping operations have undergone dramatic paradigmatic changes. Cold War period missions were characteristically termed "traditional" and "first generation" peacekeeping involving the deployment of lightly armed noninterventionists. Non-provocative UN military observers were used, or an interposition of peacekeeping forces was used, in what was termed "preventive diplomacy," deployed after the cessation of an interstate conflict to oversee and assist with the implementation of peace agreements and to stand as a buffer between hostile parties. Missions were limited to interdiction between conflict parties and did not generally allow for assertive actions even in the face of human rights violations. Traditional peacekeeping was originally designed for interstate conflicts, but was adapted to domestic conflict situations, first in the tragedy of the Congo beginning in 1960. Traditional peacekeeping missions are authorized under chapter "Six and a half" of the UN Charter to establish and police buffer zones to separate and assist in the demobilization of disputants. Their missions are premised on permission of the host state(s) and a modicum of cooperation from conflicting parties. This gives parties a reassuring way of leaving each other alone and of course, the methods used are inherently non-offensive. The force size and resources are small and lightly armed and the rules of engagement permit them to use force only in self-defense. Traditional peacekeeping is noncoercive and the deployment is not designed to restore order or stop fighting between the belligerents, unlike robust and peace enforcement missions. Traditional missions therefore lack both an offensive mission and the capacity for it. Goulding subdivides traditional peacekeeping to include unarmed military observer groups early on with the United Nations Truce Supervision Organization (UNTSO) along Israel borders with Arab neighbors, and subsequently in places such as the Western Sahara, and armed infantry-based forces deployed between warring parties as in Suez, Cyprus, Syria, southern Lebanon, and Croatia.[117] Observer missions, as distinct from other traditional peacekeeping missions, are composed of a small number of unarmed personnel to observe a peace agreement between warring states, exemplified by the United Nations Truce Supervision Organization (UNTSO) deployed to supervise the truce between Israel and its Arab neighbors in 1948.

With the end of the Cold War and reduced superpower constraints, there was a shift from "traditional" to "second generation" or multidimensional peacekeeping that was more complex and dealt more with intrastate conflicts. Multidimensional missions have extensive civilian functions supporting traditional peacekeeping forces as in the United

Nations Transitional Authority in Cambodia (UNTAC). The changes in the dynamics of conflicts in the post–Cold War era precipitated a dramatic shift in the United Nations' approach to peacekeeping, transforming it from "first generation" to "second generation" missions. In this era, conflicts are internal and protagonists include not easily identifiable, non-state organized actors, not parties to human rights and international humanitarian laws who often have disregard for rules of war. Casualties of war are now predominantly civilians who are not only the victims of war but also the targets, with mass human rights violations such as murder, rape, ethnic cleansing, and other acts of aggression targeting civilians. Missions are assigned such new tasks as varied as supervising elections, reorganizing polices forces, constructing civil administration, repatriating refugees, protecting humanitarian assistance, and other peacebuilding activities.[118] These two generations of peacekeeping are characterized by the principles of consent, neutrality, and use of force only in self-defense.

Despite this transformation, the United Nations still faces insurmountable challenges protecting civilians from human rights violations in civil wars. This led to critical scrutiny of the traditional peacekeeping operational paradigm and the search for the appropriate operational doctrine to protect civilians from the marauding attacks of conflict spoilers. Following setbacks in operations in Somalia, Bosnia, and the disastrous experiences in Rwanda, Secretary-General Kofi Annan and other prominent global policy makers recommended the doctrine of robust peacekeeping. The Brahimi Report (2000) also recommended a more assertive and deterrent UN mission force to protect civilians. These recommendations attest to the enormity of the challenges posed to traditional peacekeeping's ability to protect civilians in the post–Cold War turbulent civil war environment. Robust peacekeeping is a strong and forceful peacekeeping force, though consent-based, that one hopes would deter would-be violators of civilians' human rights as in the case of United Nations Mission in the Sierra Leone (UNAMSIL) in 1999, the first UN robust mission. The force strength is large; for example, 18,000 troops were deployed for MONUC operation, and also heavily armed with offensive weapons including armored tanks and combat helicopters. It is a mission short of full enforcement dubbed "Chapter six and three quarters" or "gray area peacekeeping" authorized under Chapter VII[119] as in MONUC that authorizes MONUC forces to use force to protect civilians. In the next section, I discuss the constituent elements or principles of traditional peacekeeping that have been the focus of discussions in the search for a more assertive mission with capabilities to protect civilians from human rights violations when deployed in violent civil wars.

Principles of Traditional Peacekeeping

Neutrality or impartiality

Neutrality or impartiality is a principle of traditional UN peacekeeping missions designed to facilitate constructive intervention and forge a solution that meets the interest of conflicting parties.[120] The purpose of impartiality is to demonstrate that the United Nations is an honest broker of peace. Traditional peacekeepers are impartial and do not advance the interest of any of the parties over the others in order to win the trust of conflicting parties and facilitate conflict resolution. When peacekeepers lose their impartiality, they risk losing that trust with serious consequences including noncooperation and attacks against them and the civilian populace. Partiality by the UN forces will therefore imperil the attainment of mission objectives. Indeed neutrality goes beyond the actions of the force to concerns about the troop's composition, excluding members with interest in the conflict. The force composition of traditional peacekeeping normally excluded major powers and states with interest in the conflict. An additional reason to exclude the major powers was to avoid bringing Cold War competition and discord into the already tenuous conflict situation meant to be pacified.

The strict definition and applicability of impartiality seems counterproductive to peacekeepers' objective of civilian protection in contemporary violent civil wars. A new conception of impartiality has evolved to meet contemporary threats to civilians. Impartiality does not mean impartiality between the belligerents, but impartiality in carrying out the UN Security Council decisions.[121] It does not mean that peacekeeping forces cannot enforce their mandate without prejudice. Rather, peacekeepers can use force against spoilers who violate the mission's mandate.[122] The basic argument is that impartiality is not a hindrance to mission objective. Absolute impartiality by peacekeepers sometimes implies indirect connivance with spoilers whose behavior is being censored by the international community, exposing UN forces to the accusation of condoning that behavior. Impartiality is a justifiable option once it enhances mission success, but it is counterproductive if its adherence undermines the achievement of the overall objective of the mission. Despite the emphasis on impartiality, United Nations' actions during the Suez and Congo crises were seen by some states as advantaging or facilitating one party against another. In the Suez crisis, the UN Security Council condemned the actions of Britain and France.[123] Similarly, United Nations Operation in the Congo (ONUC) from 1960 to 1964 initially supported the forces of Congolese prime minister Patrice Lumumba in opposition to the Katangan president

Moise Tsombe, but subsequently the Soviet Union accused the mission of favoring Western interests in undertaking military actions supposedly to keep the peace.

Consent

UN peacekeeping missions are deployed with the consent of the host country. Consent underscores the United Nations' reverence for the principle of sovereignty and recognition of the legitimacy of the local government's authority, a point important to the United Nations' continued acceptance and viability, and an assurance to respect the host countries' laws and customs. Deployment without the host country's consent may precipitate attacks on forces and civilians. In traditional peacekeeping, consent is a criterion for mission success because it is necessary for trust building and peaceful negotiations. The principle of consent makes peacekeeping less threatening and more acceptable to conflict parties and may also reduce the risk of combat casualties to mission troops. In his 1992 Agenda For Peace, Boutros-Ghali defines peacekeeping as "the deployment of a United Nations forces in a field, hitherto with the consent of all the parties concerned."[124] The absence of consent is a guarantee of noncooperation as evidenced in Bosnia where Bosnian Serbs disregarded and attacked the safe area and in Somalia where UN forces suffered a high rate of casualties in 1993. The problem is consent is sometimes unreliable and subject to manipulation by parties placing restrictions on the UN forces. For example, in the First United Nations Emergency Force (UNEF1), Egyptian president Gamal Abdel Nasser withdrew his consent, asking the UN forces to vacate Egyptian territory just before the 1967 war. It is arguable that the absence of consent risks forfeiture of parties' cooperation; however, it is also a fact that consent is not synonymous with cooperation to fulfill the mandate. For example, the Khmer Rouge and the Phnom Penh regime consented to United Nations Transitional Authority in Cambodia (UNTAC), but they refused to cooperate with several aspects of the mandate as evidenced by the Khmer Rouge's refusal to disarm.

Limited Use of Force

Traditional peacekeeping forces are lightly armed and use force only in self-defense. This is because peacekeeping forces are not designed to alter the prevailing distribution of power, neither should they appear to threaten the disputants or the local population. The use of force other than for self-defense fails to reconcile the logical and practical incompatibility between the roles of peacekeepers to keep peace and the excessive use of force as in enforcement action. Indeed there is a contradiction in the notion of "fighting for peace" inherent in such theories as collective security.[125] The use of force entails certain opportunity costs, including reduced ability to make the peace. The use of force as in robust peacekeeping inevitably undermines the forces' impartiality and downgrades its credibility and trustworthiness.[126] The loss of neutrality due to the use of force makes mission forces vulnerable to attack and increases reprisals against civilians by combatants as evidenced in the Somalia mission and in the United Nations Protection Force (UNPROFOR) in the former Yugoslavia. In the United Nations Operation in Somalia II (UNOSOM II), the employment of what was seen as excessive force converted US forces into a perceived hostile faction to the conflict, detrimental to the principle of neutrality, and subjected them to attacks by Somali factions resulting in mission failure.[127] In Bosnia, the use of force by the North Atlantic Treaty Organization (NATO) forces led to the capture of UN forces and reprisals against civilians.

Contrary, however, to the traditional notion that limited use of force is a criterion for mission success, the absence of force sometimes limits mission effectiveness especially in civil conflict zones where conflict parties use force to achieve their objective. The United Nations Assisted Mission in Rwanda (UNAMIR) and UNPROFOR missions failed because UN forces failed to use appropriate force to stop civilian killings.[128] In a crisis situation that demands the stoppage of atrocities, such as genocide, the deployment of nonlethal peacekeeping forces using force only for self-defense seems inapplicable, impractical, and inhumane as in the tragedy of Rwandan genocide and Bosnia massacres. In the contemporary, volatile, civil conflict environment involving some conflict parties that totally disregard international humanitarian laws, the deployment of forces under the principles of consensual and limited use of force seems an inadequate response to the level of humanitarian crisis in the afflicted state.[129] The 2000 Brahimi Report reaffirms this position when it acknowledges the principles of traditional peacekeeping but also observes that the United Nations should be prepared to confront lingering forces of war and violence.[130]

ONUC, for example, was guided by the principles of consent, neutrality, and limited use of force in the Congo, but it experienced significant problems. It was only after the employment of coercive measures involving offensive use of force that order was restored. Thus ONUC failed using the traditional approach, but was successful at least in its immediate mission using force not only for self-defense but also to enforce peace. Similarly, when the UNAMSIL was initially deployed under a traditional operational paradigm, it could not prevent attacks on and killing of civilians, as well as attacks on its members, by the rebel group, the Revolutionary

United Front (RUF), because it lacked effective deterrence capacity. However, when it was later handed a robust mandate, UNAMSIL provided security countrywide and protected civilians from RUF's atrocities.

UN Peacekeeping and Civilian Protection

Civil wars over the last two decades have been replete with systematic violence and mass atrocities against vulnerable civilians with increased demand on the United Nations, particularly UN peacekeeping forces deployed in those conflicts, to prevent and halt such violations. Humanitarian provisions of the laws of armed conflict and international human rights laws and norms such as responsibility to protect constitute the fundamental basis of peacekeepers' legal and moral obligations to protect civilians when deployed in civil wars. The UN peacekeeping forces' protection of civilians fulfills the United Nations' objective of saving succeeding generations from the scourge of war.[131] To meet the challenges posed by recurrent human rights violations in conflict theaters, there is the need for peacekeeping operations to contribute to larger values such as global peace, justice, and the reduction of human suffering.[132] UN peacekeepers' protection of civilians correlates with the legitimacy and credibility of peacekeeping missions. Vulnerable local civilians in civil wars always have high expectations of deployed UN peacekeepers to address violations and protect them in line with the purposes of the United Nations. Peacekeeping missions risk losing credibility and legitimacy if they fail to address civilian insecurity that adversely affects their capacity to fulfill other mandated tasks.[133] Ultimately UN peacekeeping missions, that lack the capacity to protect civilians from atrocities, such as in Rwanda, the Balkans, the Democratic Republic of Congo, and Darfur, risk mission failure. Failures in these missions to protect civilians under imminent threat damaged the United Nations' reputation and raised doubts if its commitment to its espoused ideals of human rights, peace, and security are mere rhetoric. "Such doubts cut to the core of the UN's raison d'être—'to save succeeding generations from the scourge of war.'"[134]

During Cold War traditional peacekeeping, civilian protection was not the explicit objective of UN missions, and civilian protection was intrinsically achieved through peacekeepers' strategic interposition as a buffer zone between conflict parties.[135] The prime focus of traditional peacekeeping was to prevent major powers' involvement in interstate conflicts to avoid potential escalation of these conflicts into major wars and civilian protection played second fiddle to that. The viability of strategic inhibition was a derivative of the designation of traditional peacekeeping that focused on conflicts involving state actors by imposing territorial and international legal restrictions that hindered easy accessibility to opponents' territories to harm their civilian populace. Peacekeepers' fixation on the principles of impartiality and neutrality coupled with the fear of reprisals against them by a belligerent that would misconstrue the mission forces' protection of civilians as siding with their opponents undermined their desire to protect civilians. During ONUC, UN Headquarters refused Ghanaian peacekeepers' request to take President Patrice Lumumba into protective custody with the reason that it would breach the principle of impartiality. That ultimately resulted in his murder by his opponents.[136] This intrinsic stratagem of civilian protection has proved futile in post–Cold War intrastate conflicts because they involve irregular conflict parties who are not signatories to international humanitarian and human rights laws, and who operate freely without geographic restrictions. This creates difficulty in differentiating them from the civilians that need protection. The United Nations reports on missions detailing their failure to protect civilians, which accordingly garners discussions on the United Nations' relevance in achieving global peace and security.[137] The 1999 Srebrenica report called for more effective and decisive defense of civilians and response to humanitarian tragedies.[138] Following these reports, Secretary-General Kofi Annan acknowledged the necessity of civilian protection by peacekeepers deployed in civil wars[139] and recommended robust peacekeeping as the most viable operational paradigm to protect civilians. To salvage the image of the organization, he convened on March 7, 2000, a panel on United Nations Peace Operations headed by Algeria's Brahimi, with the mandate to make "frank, specific and realistic recommendations for change in peacekeeping missions."[140] The Brahimi report acknowledged peacekeepers' responsibility to protect civilians under imminent threat using all means, including force.[141] This book examines the effectiveness of UN robust peacekeeping in staving off civilian killings when deployed in violent civil wars using both statistical models and cases. The core argument of this book is that robust peacekeeping, particularly the core elements, are the panacea to civilian killings. The next section looks at the general scholarly literature on UN peacekeeping success.

Scholarly Literature on United Nations' Peacekeeping Success

Analysts differ in perspective on the evaluative standards and common working definitions of peacekeeping "success" or "failure." There are controversies over the agreement on a common definition of peacekeeping success or failure.[142] The ambiguity of success and failure and the time framework used to measure results makes it difficult to gauge success or failure.[143] Analysts also differ on the appropriate framework for analysis and standard of evaluating success of peacekeeping missions. Diehl suggests the criterion of reducing armed conflict and contributing to conflict resolution.[144] Johansen, however, criticized Diehl's criteria claiming that it misrepresented the "real purpose of peacekeeping" and missed many of the main benefits of these missions.[145] Johansen argues that because peacekeeping missions are often not designed to address these issues, Diehl's criteria are unsuitable. Johansen stresses the need for qualitative criteria and the contribution of peacekeeping to larger values such as world peace, justice, and the reduction of human suffering. Some scholars compare mission outcome with mandates[146] and mandate completion as measures of success or failure but this approach is criticized because mandates are often vague, complex, and change during missions., and so designed to adapt to changes during missions. Other suggested criteria include whether the mission was efficiently planned and resourced, or that missions should be evaluated in reference to the constraints they operate under in respect of funding, material, personnel, and recalcitrance of the protagonists, and whether a long- or short-term perspective should be applied. Others emphasize the need for qualitative criteria and the contribution of peacekeeping to conflict management and the larger global values such as world peace, justice, and the reduction of human suffering.[147] This perspective considers peacekeeping as a tool for controlling and mitigating conflicts and defines mission success in terms of its ability to create an enabling environment for conflict resolution. Following this perspective, the roles of peacekeeping are defined as follows: "a controlled impasse";[148] "the containment of war so that others can search for peace in stable conditions";[149] "a barrier against . . . unwanted war";[150] "a deterrent";[151] "the limitation of armed conflict";[152] and "reduction of conflict casualties."[153] The commentators agree that both short- and long-term considerations are important.

Different research findings have been reached in the field using either case or empirical studies. Early peacekeeping studies were predominantly more theoretical and descriptive of case studies than empirical studies due to the problem of small number of cases and number of variables ratio.[154] The case studies comparatively analyze missions to determine the causalities of success or failure in order to learn lessons from previous missions and formulate effective policies for future operations. The problem with this approach is that different definitions, observations, and variables are used that make generalizations difficult. The initial systematic analysis focuses on the ability of peacekeeping operations to prevent the renewal of armed conflict. William Durch mentions consent, the support of the Great Powers, especially the United States, and prior alteration of the local parties' basic objectives from winning to salvaging something as important factors to peacekeeping success and confirms the findings with a study of all cases of UN peacekeeping operations since World War II.[155] Diehl identifies the characteristics of the peacekeeping operation, authorization, and political and military contexts as factors for peacekeeping success. He comparatively analyzes six cases of traditional peacekeeping and finds that third-party states, subnational groups, poor geographic conditions, and the absence of neutrality undermine confidence in the operation and engender escalation of minor crises to major ones. Diehl, however, finds that a clear mandate, financing, a command structure, the primary disputants, and the role of the superpowers have relatively minor impacts on the peacekeeping outcomes.[156] Assis Malaquias cites the importance of adequate resources.[157] The above studies focus on interstate conflicts.

Focusing on intrastate conflict, some scholars find the support of the members of the Security Council to be important to mission success[158] while other case studies focus on mission type and mission success. Some researchers argue that more "robust" and ambitious missions and peace enforcement in the post–Cold War era are inevitable, and more effective than traditional peacekeeping.[159] Donald Daniel et al. comparatively examine the missions in Bosnia, Somalia, Rwanda, and Haiti, and argues that coercive inducement is useful to bridge the gap (gray area) between traditional peacekeeping and enforcement action.[160] Others, however, argue against missions differing from traditional peacekeeping, especially the engagement of peacekeeping forces in full-scale military operations against spoilers since that would rather escalate the conflict and increase reprisals against civilians.[161] Traditional peacekeeping operations were found to obtain relative success while enforcement actions did not. A very important finding is that the UN system is inadequate to control peacekeeping operations beyond self-defense and suggests that robust or ambitious missions be taken by military alliance or coalitions led by major military powers.[162] These studies, however, have not been subjected to empirical testing to permit generalizations.

Early empirical studies focused on mission types and success in interstate war.[163] Empirical research on peacekeeping effectiveness in intrastate conflict is also recent. Virginia Fortna, in a study of the impact of peacekeeping on peace duration finds that peace, in general, lasts longer when peacekeepers are deployed than when combatants leave on their own. She, however, shows that enforcement missions have not been effective in keeping peace and may be detrimental to durable peace. In contrast to Fortna who examines interstate wars, Michael Doyle et al. focus on civil wars for the period of 1944–1997 and empirically analyze the effectiveness of UN peacekeeping operations on peacebuilding after the end of the war. They find that UN enforcement and observer missions improve the chances of peace. They also find that multidimensional peacekeeping operations reduce violence as well as improve the prospects of durable peace. They, however, report that traditional peacekeeping has no effect on ending both the violence and peacebuilding success.[164] According to Caroline Hartzell et al., third-party intervention (including peacekeeping) significantly and substantially increase peace duration[165] but Dubey, using Doyle et al. dataset but with more sophisticated statistical techniques, however, finds that third-party peacekeeping interventions have no significant effect on peace duration.[166] George Downs et al. in their study categorize the cases into degrees of difficulty and illustrate that the conflict difficulty, the degree of the effort of the international community, and outcome are related to one another. Using variables related to the conflict environment and international commitment, they find the presence of spoilers, neighboring states opposing the peace process, and valuable commodities as environmental factors for mission failures. The extent to which the major powers define a particular conflict as to their national security interest is identified as the most important international commitment factor.[167]

The findings from the studies are mixed and even contradictory. This is basically the sequel of using different definitions, variables, and observations of peacekeeping. Recent studies, however, point out that the differential findings might be due to the problem of selection effects that need to be controlled before a proper assessment of mission success is made.[168] The book uses "success" in the sense of limitation of human rights violations, specifically, the reduction in the deliberate killing of civilians, following the deployment of a peacekeeping mission. The reason for this perspective is that it is a salient primary objective of peacekeeping missions. Other goals can be successfully pursued if peace is secured. It is therefore most appropriate that such causalities be addressed. In this study, success is defined as a decrease in intentional civilian killings by combatants, which implies the effectiveness of UN peacekeeping operations during civil wars. This book focuses on the reduction in intentional civilian killings as a measure of mission success because serious political negotiations cannot be carried out in the midst of atrocities. Parties suffering costs in terms of lives will not cooperate until the killings of their people have stopped. Negotiations are possible only when mission forces are able to stave off atrocities and create an enabling environment for political negotiations.

Factors Limiting Successful Peacekeeping Missions

UN peacekeeping missions have occasioned some successes and some notable failures of collective action, but there is a continued call for peacekeeping that tends to outrun the availability of ready multinational forces. Mandates of peacekeeping forces in current operational theatres such as the Sudan, Democratic Republic of Congo, Burundi, Haiti, and Cote d'Ivoire include the protection, monitoring, investigation, and prevention of human rights violations. These functions remain integral to the United Nations in maintaining international peace and security. Human rights violations evidently have continued even in the midst of international peacekeeping, and sometimes, unfortunately, even involve the peacekeepers themselves. Human rights protection today is one of the most challenging missions undertaken by multilateral forces, since forces often intervene in chaotic situations in failed, corrupt, or crime-ridden states. The United Nations has found itself unable to successfully sustain the increasing demands placed on it. UN forces are frequently unable to stave off human rights violations. Haitians remained mired in human rights crises despite the UN peacekeeping presence. There is a dramatic shift from an initial unfounded enthusiasm to a more recent unwarranted disillusionment. Since then, the efficacy of peacekeeping as a conflict management mechanism has attracted scholarly and policy makers' concerns and has been subjected to critical, microscopic scrutiny. Thus, there remain questions about the sufficiency of the current operational characteristics, especially both the mandates and the resources of a peacekeeping force.

The all too frequent inability of UN peacekeepers to respond effectively to human rights violations of vulnerable civilians, such as the Rwandan genocide, calls for an in-depth study of certain specific capabilities, strengths, and deficiencies of UN peacekeeping missions. This is what the book sets out to do. A successful peacekeeping mission depends on both political and military factors, as well as the social environment and severity of the conflict into which forces are sent. While unilateral international interventions have been subjected to relatively detailed analysis for adequacy and success or failure correlates, there has been relatively little systematic comparative study of key factors accounting for United Nations' peacekeeping success or failure, particularly in the growing human rights domain. In the first instance, peacekeeping efforts seem frequently to be undermined or challenged by limited resources, shifting or inadequate mandates, restricted rules of engagement, insufficient troop strength or training, cultural challenges to contingent composition, and geopolitical factors. UN peacekeeping lessons learnt and General Secretary reports have noted certain recurrent difficulties. The scope and clarity of the peacekeeping mandate, either in the UN Security Council or regional organizations' authorizing resolution, may lead to failures if the scope is limited or the terms are vague. The key aspects of the mandate are its clarity and specificity. Brahimi's report emphasizes the importance of "clear, credible and achievable mandates." Mandates must be feasible and unambiguous. A clear and detailed mandate reduces disagreements over varying interpretations of mission and assists in the implementation of a successful peacekeeping strategy, minimizing cross-national or cross-contingent disagreements on how purposes are defined and how best to operationalize them.

Even though mandates have far exceeded the traditional supervision of truces and the separation of antagonists to include protection of human rights, the rules of engagement governing peacekeeping operations limit peacekeepers' capabilities to protect those rights. The amended Geneva Convention categorizes UN troops, such as the Red Cross, as noncombatants and this reflects the original noncombatant function of peacekeepers (First Protocol 1977, Additional to the Geneva Convention). Peacekeeping forces generally are limited to use force only in self-defense. Unless attacked, traditional peacekeepers may not use firepower to deter violators even though volatile situations in the field demand the necessity to stretch definitions of self-defense and the conditions for the use of force. Intrastate conflicts differ greatly from interstate conflicts around which much of international collective security system was designed. For example, in interstate conflict, conflict parties have legal obligations to observe international humanitarian laws and human rights, but in intrastate conflicts, belligerents include nonstate actors, not signatories to human rights treaties, who ignore international legal obligations in repressing population elements.

A further limitation is peacekeepers' guiding principle of neutrality/impartiality that prohibits them from altering the existing distribution of power. The concept of neutrality also covers the composition and by implication the cultural balance of forces deployed. The inclusion of countries with vested interests in the conflict jeopardizes its neutrality and the cooperation of the dissatisfied conflict party. Having ethnic "enemies" of certain conflicting parties inside the peacekeeping ranks also carries the risk of undermining the trust and acceptance of the intervening forces. Thus, there is controversy over how ethnically "balanced" peacekeeping contingents must be, a difficult task especially given the relative shortage of forces to go around in global conflicts. Clearly, neutrality and trust building in force composition and action improve the chances for acceptance and cooperation by parties to the conflict as well as the local populations in towns and villages. In reference to ethnic composition, linguistic differences may result in misinterpretations, disagreements, and noncooperation among forces diminishing the effectiveness of the peacekeeping missions. It is, therefore, assumed that the greater the cultural and the linguistic difference among contingents, the less likely the success of the mission.

The successful conduct of peacekeeping operations is bolstered by the credibility, integrity, and competence of the forces executing the mission. Thus, one would presume that more forceful and more offensively mandated operations would be more effective or successful than more traditional and restrained forms. Peacekeeping efforts, however, may be undermined by discrepancy between mandates and resources. Financial support of the operation, for example, might not match the complex mandates authorized by the Security Council. Generally, peacekeepers are only lightly armed for self-defense and financing for missions always seems to lag. Troops may be poorly trained and led, and might themselves become parties to the dispute, in worst cases becoming victimizers of the local population and indulging in crime and exploitation.[169] The international community needs to know when such breakdowns of authority and trust can be expected to occur, and what preventive measures can best preclude them whether in robust or non-robust operations.

Of equal concern is force size and strength. Even if granted the power to use force to deter violators, the size of troop contingents deployed may be insufficient. Force too quickly applied can be quickly outnumbered, which is another counterproductive outcome. Thus, it is important to explore the relationship between troop deployment size and the other factors mentioned in conditioning positive or negative peacekeeping outcomes, depending on mission; for example, the betterment of human rights conditions. It is reasonable to question whether the number of troops deployed influence mission success. Of presumed importance is the ratio of peacekeepers to the general population and most importantly to combatants. The US expeditionary force in Haiti, as one case, numbered 20,000, outweighed the local Haitian forces by three to one. We can therefore assume that the ability of the US forces to curtail human rights violation was rather high, especially given the technological disparities between the armies.[170] Yet even in Haiti, human rights and social conditions may not have been improved for the long term by the relatively limited scope or perhaps limited interest of the US mission and its subsequent replacement by the Organization of American States (OAS). The geographic area and topography of the host country in which forces are deployed are of considerable importance in considering mission strength. The number of combatants in relation to area and to peacekeeper density may provide insights into the prospects for peacekeeping mission successes. The larger the areas of deployment, the more difficulties peacekeepers face in achieving their objective. Small-sized troop contingents such as those of the African Union (AU) in Darfur make it impossible to monitor thousands of kilometers, offering opportunities for aggressors or "spoilers" to violate ceasefires and human rights. According to Baccock, peacekeeper density for successful missions is an average of 1,052 peacekeepers per sq. km whereas that for unsuccessful missions is 0.0500 peacekeepers per sq. km. The topography of the area also affects peacekeeping forces' ability to monitor human rights. Operations in an area of natural barriers such as mountains and swamps affect operational success, especially as local forces might have an advantage of knowledge of and maneuverability in such terrains. Low population density offsets problems of detecting violations in large areas, but high population density in relation to small force strength limits ability to detect violations.

The United Nations also maintains troops as national units coordinated by an overall UN command. Various problems may arise from differences in command rules, culture, language, training, equipment operability, and operational and conceptual definitions. A possible difficulty from these differences is integrating these various national units with differential resources, training and operational concepts into a coordinated and comprehensive unit. A well-trained and fully integrated multinational force with common understanding of operational concepts facilitates cohesion and successful execution of mission.

In an era of collective efforts to stave off human rights violations, the United Nations will need to reexamine its guiding principles, internal and operational characteristics of peacekeeping to address the hindrances to peacekeepers' contribution to the United Nations' pursuit of global peace and protection of civilians from human rights violations. This book proposes to assist in doing so by systematically analyzing the varied factors underlying success and failure of missions, particularly in terms of human rights observance. In this book, I borrow from the rational choice theory of conflict termination to argue that mission success is attained if the mission has the capability to alter the calculations of conflict parties to choose peaceful negotiations instead of continued fighting. Following the logic from the costs of war termination theories, I develop the *formidable barrier* model of UN peacekeeping mission success. Both the rational choice theories of war termination and *formidable barrier* model of peacekeeping success are discussed in detail in Chapter 3.

3

Robust Peacekeeping and Civilian Protection

Civil wars over the last two decades have been replete with incidents of systematic violence and mass atrocities against vulnerable civilians with a heightened demand on the United Nations, particularly on the UN peacekeeping forces deployed in those conflicts, to prevent and halt such violations. The protection and advancement of human rights is a fundamental objective of the United Nations in its avowed commitment "to save future generations from the scourge of war"[1] and peacekeeping is one of the most innovative mechanisms to achieve this objective. Humanitarian provisions of the laws of armed conflict and international human rights laws and norms such as "Responsibility to Protect" (R2P) constitute the fundamental basis of peacekeepers' legal and moral obligations to protect civilians when deployed in civil wars. The UN peacekeeping forces' protection of civilians fulfills its objective of saving succeeding generations from the scourge of war.[2] Despite peacekeeping missions' awareness of their responsibility to protect civilians under imminent threat, we continue to witness systematic violence and egregious human rights violations against civilians even when UN peacekeepers are deployed in violent conflicts as in the case of UNAMIR during the 1994 Rwandan genocide.

The United Nations' traumatic failures in the 1990s, when it failed to protect civilians from genocide in Rwanda, crimes against humanity in the former Yugoslavia, and now the systematic use of rape as a weapon of war in Darfur and the Democratic Republic of Congo have damaged UN reputation and raised questions if its commitment to its espoused ideals of human rights, peace, and security is mere rhetoric. The United Nations' noninterventionist traditional peacekeeping operational doctrine, which lacks requisite "muscle" and was originally designed for interstate conflicts, seems an inadequate doctrine to protect civilians in post–Cold War violent civil wars.

This necessitated the reassessment of the fundamental principles underlying UN peacekeeping doctrine to enable UN peacekeepers to take a much more active role in the protection of civilians when deployed in violent civil conflicts.[3] A robust peacekeeping mission—a mission with adequate deterrence capacity to stave off civilian killings—emerged as the appropriate mechanism to enable UN peacekeepers protect civilians under imminent threat. Today, expectations have steadily risen and UN peacekeeping operations have been given Chapter VII robust mandates that routinely include authorization for peacekeepers to take measures to protect civilians under imminent threat of violence.

Some studies of third-party intervention suggest that the key to any intervention strategy is to alter the calculations in a manner that leads to the outcome preferred by the intervener.[4] The primary objective of the UN peacekeeping missions, as a form of third-party intervention, is to keep peace. Thus a modicum of peace must be established and preserved. In this sense, the key to peacekeeping success is its capacity to alter combatants' calculations by raising the costs of continued or renewed fighting and benefits of choosing and persisting with peaceful relations. Partly, the scope of such calculations will be beyond the scope of troop deployment, since parties will calculate their costs and benefits politically, based on the situation and their interests and capabilities. Peacekeeping missions, however, can affect such calculations both in cost/benefit terms and in helping parties save face, resolve prickly on-the-spot issues and irritations, and engage the interest and support of the wider international community.

This chapter discusses civilian protection and the evolving ideas of "robust" peacekeeping. The chapter also presents the concept of the *formidable barrier* model of robust peacekeeping success derived, in regard to the cost/benefit calculation, from the rational choice theory of conflict termination. The model posits that the success of peacekeeping missions in staving off deliberate civilian killings is a function of its deterrent capacity. Robust peacekeeping, by designation, it is argued, has the ability to provide the necessary barrier or impediment to intentional civilian killings by combatants by imposing some military constraints on the belligerents.

The rest of this chapter is organized as follows. The next five sections discuss the historic development, theory, rationale, elements, and dilemmas of robust peacekeeping. The seventh section presents the rational choice theory of conflict termination. The eighth section sets forth a model of peacekeeping success, the *formidable barrier* model—based on rational choice theory, using a game theory model. The final section introduces a set of hypotheses that will be tested in chapters 5 and 6.

Robust Peacekeeping and Civilian Protection—Historical Development

The United Nations proclaims in the preamble of its Charter the objective of "saving succeeding generations from the scourge of war."[5] The UN peacekeeping forces' protection of civilians under imminent threat in civil wars fulfills this objective. However, over the last two decades, the world has witnessed devastative armed conflicts characterized by systematic violence and conscience-shaking atrocities against civilians. The international community cried out to the United Nations and to UN peacekeeping missions in particular to stave off or halt such crimes. Peacekeepers' obligations to protect civilians derive from humanitarian provisions of the laws of armed conflict and international human rights laws and norms clearly stated in major human rights instruments. These international provisions and norms constitute the fundamental basis of peacekeepers' legal and moral obligations to protect civilians when deployed in civil wars. The principles of human rights and concern for human dignity and rule of law legally and morally oblige UN mission forces to act using some level of force to protect human rights rather than stand by when the most shocking, egregious violations are committed against defenseless people who the UN Charter proclaim to protect. UN peacekeepers' objective of protecting civilians sustains the legitimacy and credibility of the organization's peacekeeping missions. However, UN peacekeeping missions' capacity to protect civilians in violent civil wars has been hampered by its strict adherence to traditional principles of peacekeeping and neorealist considerations of the five permanent members of the Security Council whose national interests, rather than the severity of threats to civilians, affect the authorization and deployment of robust missions. Cold War ideological antagonism and disagreements, which for decades generated distrust and hostility among the major power blocs, have paralyzed the Security Council and crippled the missions' abilities to respond to egregious crimes and protect civilians.

Civilian protection during Cold War peacekeeping, though provided, was not an explicit objective of missions as it was regarded as intrinsic to the nature of peacekeeping. Protection of civilians was to be achieved through the use of inter-positional strategies.[6] The primacy of peacekeeping at that time was the management of tensions between major powers with the focus on constraining the potential escalation of disputes into interstate wars. Peacekeeping missions' fixation on the principles of impartiality and neutrality coupled with the fear of being accused of partiality and consequential reprisals against them by conflict parties, constrained their capacities to protect civilians under imminent threat. For example, the UN headquarters, during its mission in the Congo (ONUC-1960-64), refused Ghanaian peacekeepers' request to take Prime Minister Patrice Lumumba into protective custody on the ground that it would breach the principle of impartiality[7], even when there was an imminent threat to his life. The consequence of this was that he was eventually murdered by the rebels.[8] Post–Cold War era minimal antagonism within the Security Council revived an optimism about the potential of the UN peacekeeping missions to respond to global security threats suitably and protect civilians under threat. The collapse of the Cold War ideological barrier and distrust and hostility among the major power blocs aroused new hopes and possibilities to successfully meet threats to international peace and security. This led President G. H. W. Bush to proclaim a new world order of global peace. However, the massacres in Rwanda and Srebrenica in the early mid-1990s in the presence of UN mission forces at the time when human rights protection had assumed primacy in UN peace operational doctrine dimmed the optimism and confidence in the United Nations' abilities to protect civilians under imminent threat by spoilers. UN peacekeeping missions of the mid-1990s notoriously had negative images associated with failure and moral cowardice in the face of evil.[9] A general sense of crisis, failure, and hopelessness gained currency, and civilian protection overshadowed subsequent debates on the future of the United Nations.[10] A theme central to the debates after Rwanda and Srebrenica massacres was the United Nations' ambivalence regarding the use of force since "saving the next generation from the scourges of war"[11] or meeting global security threats paradoxically may require use of military force. Rwanda and Srebrenica ultimately prompted questions on the United Nations' operational capacities as well as the viability of its traditional operational doctrine to protect civilians leading to the demand for a more competent operational paradigm to stem human rights violations in civil wars. It ignited a fundamental reassessment of the continued relevance of traditional peacekeeping and of the adoption of robust peacekeeping as a better mechanism to stave off intentional civilian killing. Global condemnation of repeated failures of the United Nations to prevent genocide and other terrible crimes against humanity in the 1990s triggered gradual evolution from its traditional noninterventionist operational doctrine to a consciously interventionist robust operational paradigm with explicit Chapter VII authorization to protect civilians under imminent threat in violent civil conflicts. The normative doctrinal shift to robust mandates to protect civilians was fostered by certain critical conceptual developments that altered the United Nations' traditional conception of the use of force by peacekeepers for self-defense only and the eventual adoption of robust mandate to use force to protect civilians under imminent threat. The critical documents informing this strategic shift include the 1992 *Agenda for Peace* and its 1995 Supplement; former Secretary-General Kofi Annan's two influential articles in the late 1990s: *Walking the International Tightrope* and *Two Concepts of Sovereignty*; the 2000 Brahimi Report; the 2001 Responsibility to Protect; the 2004 Report of the High-Level Panel on Threats, Challenges and Changes; and the 2008 United Nations Peacekeeping Principles and Guidelines, or the *Capstone Doctrine*—all centering on the use of force to protect civilians.[12]

Reflective of the new optimism and activism in the post–Cold War era, the 1992 *Agenda for Peace,* published by then Secretary-General Boutros Boutros-Ghali, set out an optimistic and ambitious agenda for a proactive and principled United Nations focusing on human rights protection.[13] The document captured the mood of the period by challenging notions of "absolute and exclusive sovereignty." The publication pressed for greater human security and affirmed the right of the United Nations to intervene in a state under Chapter VII of its Charter to protect citizens' human rights.[14] Boutros-Ghali in the *Agenda for Peace* acknowledged that the United Nations should use force when all peaceful means had failed and that the credibility of the United Nations was at stake if it did not.[15] In spite of its optimistic and ambitious agenda of the United Nations' military humanitarianism, the document's working definition of peacekeeping was so tepid that it failed to foster the protection of civilians by peacekeepers. The United Nations' passivity toward a concrete paradigm shift to robust mandates continued until failures in Bosnia and Rwanda. Following these failures, the UN Department of Peacekeeping established in 1995 the Lesson Learned Unit (now Best Practices Unit) to examine and report on the success and failures of its missions but its analysis was less than rigorous.[16] The Unit's 1996 general report, *Multidisciplinary Peacekeeping: Lessons from Recent Experience*, despite a section on "Relations with Local Population," did not critically examine how peacekeepers should respond to crimes against humanity.[17]

Secretary-General Kofi Annan played a central role in igniting concerns for United Nations' dismal and abysmal responses to human rights violation when he commissioned two high-level inquiries to critically investigate the United Nations' failure to react effectively to 1994 Rwandan genocide and 1995 Srebrenica fall.[18] The Report on Rwanda's genocide criticized the lack of political will by Security Council members and the United Nations in general to stave off the genocide despite credible evidence.[19] Secretary-General Annan reiterated civilian expectation of protection by peacekeepers deployed in civil wars.[20] He urged Member States to soberly reflect on United Nations' "inadequacy of symbolic deterrence in the face of a systematic campaign of violence and role of force in the pursuit of peace."[21] Secretary-General Annan's two articles, *Walking The International Tightrope* and *Two Concepts of Sovereignty,* both published in 1999, greatly impacted the debate on UN peacekeeping principles and guidelines on civilian protection.[22] In *Walking in the International Tightrope,* Annan detailed the harsh reality the organization faced at the end of the twentieth century in a world filled with conflict and hinted at the need for a paradigmatic shift to adopting the requisite mechanism to confront the menace of civil wars. He stated:

> [. . .] must deal with the world not as I would wish it to be, but as it is. I must confront it with a sense of reality about how far a leader can be pushed by peaceful means, and how long it will take to bring peace where a state of war exists.[23]

To stave off atrocities against civilians in future operations, Annan felt there is a need to balance peaceful and forceful mechanisms at the disposal of the organization. Annan, to achieve the objective of civilian protection, redefined the principle of impartiality, stating that "impartiality does not and must not mean neutrality in the face of evil."[24] Annan's conclusion was heavily impacted by the fact that UN failures in Bosnia and Rwanda occurred under his watch as head of the UN Department of Peacekeeping Operations (DPKO).[25] Annan, in *Two Concepts of Sovereignty,* acknowledged "the need for timely intervention by the international community when death and suffering are being inflicted on large numbers of people, and when the state in charge is nominally unable or unwilling to stop it."[26] Annan by these publications, promoted an interventionist operational paradigm informed by Rwandan and Bosnian failures. Being by-standers in the face of mass atrocities against civilians was no alternative, as reflected in President Clinton's "never again" atonement speech to Rwandans.

The 1999 Secretary-General Reports on the protection of civilians in armed conflicts submitted by Annan established the structural linkage between humanitarian protection strategies and civilian protection obligations of UN peacekeepers. The report affirmed the Security Council's obligation to authorize UN peacekeepers to protect civilians in conflict theatres or enforcement measures and suggested specific tasks for peacekeepers to include, among others, deterring and addressing abuses of civilian populations by the conflict parties. [27] Secretary-General Annan, in his *Millennium Report* of March 2000, also challenged UN Member States and urged them to "forge unity" on identifying the basis for humanitarian intervention and how "we respond to a Rwanda, to a Srebrenica-to gross and systematic violations of human rights that affect every precept of our common humanity."[28] Secretary-General Annan in the report acknowledged that the world could not be by-standers when heinous atrocities are committed against defenseless civilians and that the United Nations in particular has a moral obligation to protect civilians from abuses by dictatorial leaders as well as spoilers. The interventionist stance promoted by the 1999 reports and Secretary-General Annan's reports ignited debates on civilian protection in the Security Council. Human security concerns trickled into subsequent mandates of the Security Council that authorized ambitious mandates with stronger language of civilian protection in armed conflicts. In response to Secretary-General Annan's report concerning the situation in Africa and the protection of civilians in armed conflicts that highlighted the insecurity of civilians as victims and prime targets of warring parties' attacks,[29] the Security Council unanimously adopted on September 17, 1999, Security Council resolution 1265 to address it. The Security Council in resolution 1265 discussed the protection of civilians during armed conflicts.[30] Resolution 1265 among others expressed the Council's "willingness to consider how peacekeeping mandates might better address the negative impact of armed conflict on civilians."[31] Recalling Security Council resolution 1265, Security Council resolution 1296 on protection of civilians in armed conflicts affirmed resolution 1265. UNSC resolution 1296 discusses steps to enhance the protection of civilians during armed conflicts addressing the need to provide better, adequate resources to enhance peacekeepers' capacity to protect civilians under imminent threats.[32] The United Nations Mission in Sierra Leone (UNAMSIL) mandated by Security Council resolution 1270 was the first UN peacekeeping mission to be provided with an explicit, robust mandate primarily authorized under Chapter VII in 1999 inter alia "to afford protection to civilians under imminent threat of physical violence."[33]

The 2000 Brahimi Report, inspired by the United Nations reports on the Rwandan genocide and the fall of Srebrenica, also recommended more assertive and deterrent mission force to be able to confront the challenges of human rights violations in contemporary volatile conflict theaters.[34] It redefined the principle of "impartiality" to mean "adherence to the principles of the Charter and to the objectives of mandate rooted in those Charter principles."[35] The report argued that in dangerous situations with "obvious aggressors and victims," peacekeepers, "may not only be operationally justified in using force but morally compelled to do so," and should therefore be equipped with "robust rules of engagement."[36] The use of force by peacekeepers to protect civilians under imminent threat, supports the principles of the United Nations and fulfills the Security Council's responsibility to maintain international peace and security. The 2000 Brahimi Report, supported by Secretary-General Kofi Annan and other prominent global policy makers, recommended a new conception of "robust peacekeeping," defined as a strong force or mission with broadly defined protective mission mandates, to protect the population and reduce human rights violations in areas of deployment.

The normative paradigm shift to robust peacekeeping is reinforced by the publication of the international norm of "Responsibility to Protect" (R2P) in December 2001.[37] The basic idea of the principle is that state sovereignty connotes responsibility to protect its citizens from heinous crimes but when their parent states are unwilling or unable to protect them, the international community through the United Nations has a moral responsibility to intervene to protect the citizens. Also emerging from the report is the basic idea that "human beings can count more than sacrosanct sovereignty enshrined in Charter Article 2(7)"[38] with an emphasis on noninterference in domestic affairs. This doctrine was amplified by Secretary-General Annan, when he stated that "state frontiers . . . should no longer be seen as a watertight protection for war criminals or mass murders."[39] With the inception of the "Responsibility to Protect" doctrine and the concept of international humanitarian concern into the international security discourse, we observe that a blurring of domestic and international jurisdiction has taken place and the limitation imposed by Article 2(7) of the UN Charter seems to diminish. Peacekeeping operations consequently moved from the traditional interposition and ceasefire monitoring to include protection of human rights.

The United Nations' historic operational paradigm shift is also reflected in *The Report of the High-level Panel on Threats, Challenges and Change, A More Secure World: Our Shared Responsibility* of 2004, which highlighted the role of the United Nations in facing the security challenges of the future.[40] The report, informed by events in Iraq, Afghanistan, and Darfur, endorsed the acceptance of R2P as an emerging norm and recommended criteria for international military intervention.[41] The report identified five considerations when the Security Council makes decisions on the use of force: seriousness of threats, proper purpose, last resort, proportional means, and balances of consequences. It advised the Security Council to be more proactive in the future, "taking decisive action earlier" in the face of threat. The DPKO and the Department of Field Support's (DFS) 2007 *Guidelines and Principles for Peacekeeping Operations*, also known as the *Capstone Doctrine* that aims to create a set of common guidelines for UN planners and practitioners "recognizes the need for a clearer articulation of the doctrinal foundations of United Nations peacekeeping operations [. . .]."[42] It urges peace operations to use force proactively to defend their mandate and protect civilians under threat. The Capstone, however, states that force should only be used as a last resort, and with restraint, with the ultimate aim to "influence and deter spoilers working against the peace process or seeking to harm civilians; and not to seek their military defeat."[43] Since the first landmark UNAMSIL resolution 1265 (1999), most UN missions have been authorized under Chapter VII and provided with robust rules of engagement (ROE) to "take the necessary action to ensure the security and freedom of movement of its personnel and to afford protection to civilians under imminent threat of violence . . . ." [44] These include UN missions in the Democratic Republic of the Congo (1999), Kosovo (1999), East Timor (1999), Ethiopia–Eritrea (2000), Liberia (2003), Haiti (2004), Côte d'Ivoire (2004), Burundi (2005), Sudan (2005), Timor–Leste (2006), Darfur, Sudan (2007/2008), Lebanon (2007), and Chad–Central African Republic (2007).

Theory of Robust Peacekeeping

A robust peacekeeping mission is one that has the credibility or capacity to deter spoilers who threaten the mission's mandates and civilians using all necessary means, including force, to defend itself and fulfill its mandate.[45] The 2000 Brahimi Report recommends the concept of "robust" peacekeeping as part of the reforms and revitalization of peacekeeping to enhance its capacity as an instrument for the maintenance of international peace and security. According to the report, "no amount of good intentions can substitute for fundamental ability to project credible force" and it suggests that more assertive forms of intervention can mitigate ongoing conflicts or protect civilians.[46] This concept was facilitated by the end of the Cold War and by the embarrassment of unsuccessful UN experiences. It was further helped by the subsequent development of humanitarian posture within international peace, political, and legal discourses to capture the mood of contemporary security realities. Secretary-General Annan outlined the shift to "robust peacekeeping" when he recommended that the United Nations should abandon the outdated concepts of neutral peacekeeping and replace them with a more muscular form of peace operation if it is to avoid the kind of fiascos that occurred in previous missions. The Secretary-General seemed to put the premium on mission type. He acknowledged the challenges posed by spoilers in the fulfillment of mission mandates and recommended that the UN peacekeepers be equipped and given robust mandates to withstand these challenges.[47] He was convinced that all future UN missions should be able to deal with any circumstance, including if a signatory reneges on an agreement. In a similar vein, the 2005 World Summit Outcome document recommended that peacekeepers should have "adequate capacity to counter hostilities and fulfill effectively their mandates."[48] The emphatic tenor of the robust paradigm is the emphasis on a deterrence or use of force to protect civilians and fulfill mission mandates. The objective is not to defeat, but to present a credible military threat to deter, deny, neutralize, and convince spoilers that violence will not succeed.[49] The use of force is meant to stop or deter hostile or violent threats and acts from spoilers who employ murderous force to undermine the mission objective as well as achieve their own objectives. In robust peacekeeping, therefore, the rules of engagement transcend the traditional notion of self-defense to include the use of force to deter would-be spoilers and achieve humanitarian objectives.[50] The fundamental presumption is that appeasement of aggressors or violators is counterproductive to peace. Robust peacekeeping is informed by the rational choice theory of conflict termination, which postulates that the decision to fight or not is a function of the cost of continued fighting and the benefit of achieving a settlement.[51] War will occur if combatants make a calculation that the benefits of escalating a conflict to a violent confrontation outweigh the costs. On the other hand, if they perceive that the continuance of war threatens their survival and political objectives more than the abandonment of hostilities, peace will occur. Consequently, the combatants' decisions on their strategies of war and peace will be determined by a rational calculation of the constraints and opportunities that confront them at any given time. Borrowing from the rational choice theory of conflict termination, I formulate the *formidable barrier* model of UN robust peacekeeping mission success, discussed in detail below, to illustrate the viability of robust peacekeeping as the more appropriate mechanism than traditional peacekeeping to stem human rights violations. The concept of robust peacekeeping is exemplified by UN mission in Sierra Leone.

Constituent Elements of Successful Robust Peacekeeping Missions

Successful robust missions in violent civil wars with stiff opposition from spoilers require well-equipped and trained, large troop contingents, major powers included, with robust rules of engagement and the commitment to execute mission objectives. Contemporary UN missions are launched into failed states characterized by the absence of law and order, high levels of human rights violations, and deprivation and suffering among civilians.[52] In most situations, the spoilers pursue their parochial objectives, employing intimidating or terror tactics targeting the population to force people to comply with their diktats or to mobilize support for their cause or to intimidate their opponents. In this volatile operational environment characterized by nonconsensual attitudes and lethal opposition from conflict parties, the rules of engagement for traditional peacekeeping seems inadequate to deal with the realities of threat on the ground. In contrast to this, the robust paradigm permits peacekeeping forces to use force to fulfill their mandate, reflecting the humanitarian concern to guarantee the security of the civilian population under imminent threat of physical violence.[53] UN peacekeepers must be provided with "robust rules of engagement" to defend the mandate against those who seek to undermine peace accords by violence,[54] and to establish the credibility of the United Nations.[55] The United Nations' credibility was tainted when UN forces failed to use force in Rwanda to protect civilians being hacked to death. In contrast today, there are calls on the Security Council to deploy a robust force in Darfur to stop the senseless killings of vulnerable civilians there.[56] A stronger UN force in terms of sufficient numbers and equipment, requires a mandate that permits them to protect civilians using all necessary means. Without such a forceful mandate and robust rules of engagement, UN troops cannot protect civilians in contemporary turbulent civil war theatres that involve spoilers whose war stratagem is targeting civilians to seek concessions from governments. A stronger deterrent force is also essential to gain the confidence of the warring parties to monitor and guarantee a peace agreement. [57] The primary purpose of the use of force is to halt hostilities and promote stability in the conflict area.

Thus in robust peacekeeping, the principles governing "traditional peacekeeping" have been significantly altered (see table 3.4 in Appendix 3 for the characteristics of the different mission types of traditional, robust, and enforcement actions). The rules of engagement transcend the traditional notion of self-defense to include the use of force to deter and respond to spoilers' threats and protect civilians,[58] reflecting contemporary humanitarian concern in peace operations. The presence of a credible military force deters, denies, neutralizes, and convinces spoilers that violence will not succeed. The UN General Assembly acknowledges that robust forces have proven to be the best deterrent against those who mean harm in mission areas.[59] The Economic Community of West African States Monitoring Group (ECOMOG), for example, was praised by the United States for giving the United Nations "muscle" in preventing the RUF attacks against civilians in Sierra Leone. United Nations Organization Mission in the Democratic Republic of Congo (MONUC) forces also use force against spoilers that terrorized the Ituri populace. Following the Rwandan and Bosnian debacles, the protection of civilians from atrocities has assumed a central importance as a criterion for mission success. Various international norms such as the universality of human rights have extended the definition of self-defense to cover civilian protection, and the United Nations' "Responsibility to Protect" has redefined self-defense to cover civilian protection, acknowledging the primacy of the protection of civilians in contemporary volatile civil conflict zones. Gen. Romeo Dallaire, UNAMIR Force Commander, underscored the primacy of protecting civilians when he regretted the incapacitation of the United Nations in protecting the Rwandese populace from being massacred.[60] President Clinton's atonement speech to Rwandese genocide survivors for not helping to stop the genocide underscores the primacy of civilian protection in peacekeeping missions today. As Kenneth Roth, Executive Director of Human Rights Watch, stressed, the "challenge is to ensure that the United Nations troops are authorized to halt attacks on civilians, not just stand by and watch the killings continue; a stronger international force is essential to deter further attacks and protect civilians."[61]

A fundamental condition for the success of all peacekeeping missions, including robust missions, is the provision of sufficient resources including funds, weapons, and troops. Successful missions require resource–mandate compatibility. Almost all UN peacekeeping operations that fail suffer from mandate–resource discrepancy. They all face financial difficulties with operational expenses exceeding contributions due to persistent refusal of Member States to honor their commitments. In Bosnia, the UN troops were ill equipped to protect the safe areas, hence they were overrun by the Serbs. Similarly, there was a huge resource–mandate discrepancy in the Rwandan mission. UNAMIR, saddled with small light-armed personnel, could not stop the genocide and had to withdraw in the face of these atrocities. UNAMIR's Force Commander, Maj.-Gen. Romeo Dallaire, confirmed the need for greater troop strength when he argued that: "a force of 5,000 personnel rapidly deployed could have prevented the massacres in the South and West Rwanda."[62] If an outside force is to intervene successfully in a turbulent, violent war environment like the Democratic Republic of Congo, its troops must be well equipped. The need for adequate resources has captured the attention of the personnel of the United Nations. For example, Sangsnit, Thailand's UN representative, points out that having sufficient troops would increase their deterrent capacity and protect United Nations' personnel and the civilian population, as well as enhance the effectiveness and credibility of the mission itself.[63] The Rwandan UN representative Nicholas Shalita, in similar sentiments, expressed content with the adoption of UN resolution 1493, which increased the troop strength and capability of MONUC and gave it a robust mandate.[64]

In robust peacekeeping, the conception of peacekeeping has broadened from the earlier idea of lightly armed, neutral UN peacekeepers to much more activist orientations using heavy combat weapons to hunt down spoilers, similar to the MONUC mission that uses tanks, armored personnel carriers, Mi-25 attack helicopters, mortars, and rocket-propelled grenade launchers to root out the insurgents who prey on the people in

Ituri. In reference to the use of heavy arms, the Force Commander of the mission, Lt.-Gen. Babacar, stated that "it may look like war but it is peacekeeping."[65] Not only do heavy weapons make them combat ready, but also enhance their capacity to deter spoilers through the show of credible force. The show of force signals to spoilers the intentions and readiness of peacekeepers to counter violence with force and also to underscore its intolerance of and opposition to the spoilers' disruptive behaviors that undermine the mission mandate or objective. UN troops have to be fully equipped to fulfill their mandates and Secretary-General Annan acknowledges the need to rethink how we equip troops and prepare them for all eventualities.

The support and military involvement of major powers in robust peacekeeping missions is of greater relevance. Multinational forces, with participation of troops from developed countries, reflect global collective responsibility more appropriately and provide robustness. Major powers in the Cold War era were barred from participating in peacekeeping missions to guarantee neutrality and avoid bringing the Cold War into the conflict theatre. However, the changing dynamics of post–Cold War conflict, with associated lethal consequences for global security, make robust peacekeeping imperative and require well-resourced and trained troops of the developed countries take part in UN peacekeeping missions. A successful mission in an internal conflict demands the support of the five members of the Security Council, because they possess great influence over the cost of operations. In addition to giving the mission a stamp of legitimacy and seriousness, the developed countries have the critical resources, both human and material, that a credible deterrent UN peacekeeping force needs. The realization that the world community, superpowers included, supports an intervention may also exert pressure on protagonists to halt hostilities. UNAMSIL, for example, was cited as a model of robust mission because the involvement of the British troops ensured its protection of civilians from RUF attacks. Peacekeeping missions are successful when they reflect international commitment to peace and security, with developed countries "ramping up" their participation, especially those with niche capabilities.[66]

Why Robust Peacekeeping?

Robust peacekeeping is a response to the contemporary turbulent civil war environment where civilians have become the primary targets of spoilers, to protect civilians from violations by spoilers, to bridge a doctrinal gap between peacekeeping and enforcement actions, and to create the enabling environment to peacefully resolve conflicts. Civilian protection by robust missions also guarantees the legitimacy and credibility of the United Nations as well as the mission and is critical to a mission's success. Post–Cold War, turbulent civil wars necessitated a robust response in areas of UN mission force deployment in order to fulfill mandates. The post–Cold War era saw proliferations of civil wars along national, religious, and ethnic fault lines involving both state and nonstate actors. These pose challenges to global security and peace with disastrous and lethal consequences for millions of civilians. The conflicts are often continuing and characterized by absence of any peace to keep, as well as by inconsistent and ambiguous consent of the local parties, and difficulty in maintaining impartiality.[67] Peacekeeping operations today mostly respond to violence between rival factions within states that have either collapsed or are in the process of collapsing. The difficulties and threats in the operational environment raised doubts about the suitability of traditional peacekeeping in resolving these types of conflict. The principles of traditional peacekeeping seemed too idealistic to capture the realities of the post–Cold War turbulent and violent internal conflicts with lethal consequences for civilians as in Darfur. This precipitated the propositions for a change in the operational doctrine of UN traditional peacekeeping to robust peacekeeping to meet contemporary challenges posed by the changed conflict environment. Robust peacekeeping is the most appropriate mechanism in situations where peace has broken down due to the intransigence of one or more parties.[68] UN peacekeeping forces therefore do not necessarily have to obtain the consent of all parties to the conflict, nor must they use force solely in self-defense. In the new operational paradigm, UN forces are mandated among others to deter and respond to threats to civilian populations. Wole Soyinka captures this position stating that

> Can we allow the UN to be abandoned, discredited and divested of moral influence? If so, the world will see cruel panoply of violently contesting force, not only in Africa . . . . Of course this means that when the UN goes anywhere, it does not fool around with incompetent, lightly armed or underfunded troops. There is little point to future UN operation unless they will have what it takes to match fighting forces on the ground and dominate the situation.[69]

Civilian protection by UN missions also guarantees the legitimacy and credibility of both the United Nations and the mission. Robust mandates often outline the United Nations' intention to defend its mandate, personnel, humanitarian actors, and civilians under threat from physical harm by spoilers. A mission's successful protection of civilians under threat has implications for the legitimacy and credibility of any mission and of the United Nations itself. The United Nations is concerned that its credibility and legitimacy will be undermined if it does nothing while innocent civilians are killed by spoilers, especially predatory rebel forces that benefit from exploiting natural resources in a continued fighting and have no conceivable political interest in an organized government. This will include terrorist organizations. This realization is reflected in the International Commission Report 2001 titled "Responsibility to Protect" (R2P) that emphasizes the primacy of protecting human rights.[70] United Nations Secretary-General Kofi Annan amplifies the need to protect by stating that "state frontiers . . . should no longer be seen as a watertight protection for war criminals or mass murders."[71] The Brahimi Report also urges that peacekeepers should be allowed to discard their neutrality when faced with rebel groups slaughtering civilians.[72] A principal objective of "robust peacekeeping" is therefore to protect human rights. The failures of UN missions to protect civilians from human rights violations in Bosnia, Rwanda, the Democratic Republic of Congo, and Darfur damaged the United Nations' reputation and credibility, and discredited its peacekeeping doctrine. Failures to protect civilians generated questions among civilian populations in conflict theaters and the international community about the United Nations' commitment to its espoused ideals of human rights, peace, and security. It raised doubts about United Nations' commitment to save succeeding generations from the scourge of war as mere rhetoric.[73] In turbulent, complex conflicts, where civilians are targeted, the United Nations cannot afford to be discredited by failing its duty to protect civilians where peacekeeping missions are deployed. Civilian protection also correlates with mission success. The UN missions in Rwanda and Bosnia were deemed failures because of their inability to stave off mass killings despite their presence. Successful missions are those that capably protect civilians such as UNAMSIL whether included in its mandate or not. The fulfillment of other mission mandates, such as monitoring elections, hinges on the security and stability within the country. For example, successful elections can only be held in a secure and guaranteed stable environment where the people are free and safe to travel to vote. Establishing political stability and sharing of power will only succeed where stakeholders do not have to fear for their lives.[74] UNAMSIL became effective in staving off the RUF's attacks on civilians, and in guaranteeing stability, when it adopted a robust mandate.

The protection of civilians is also a critical component for a sustainable political peace. By limiting violence through its deterrent posture, robust peacekeeping may create the enabling environment to peacefully resolve conflicts. The presumption is that robust peacekeeping with its deterrent posture may prevent human rights violations bred by violence, and establish minimal compliance with human rights in a state where the rule of law has broken down, enabling political negotiations for peaceful conflict resolution. Before diplomatic efforts can be successful, violence against civilians and the cycle of impunity must be broken or interrupted. The political settlement of the conflict and the establishment and sustenance of governance, security, and the rule of law may be difficult with systematic or extreme violence against the population. In conflict management, intense conflict with its associated human rights violations undermines the prospects of mediation success. A mission's ability to ward off threats to civilian population will strengthen its ability to secure an enabling environment for a negotiated peace between the conflict parties. In the absence of nonconformity from spoilers, a measure of military force among other essential factors is necessary to deter them. Until violence is stopped, or at least managed, it is unlikely that any attempts to resolve competing interests underlying a conflict succeed. By implication, factors that lessen the intensity of conflict contribute to peacemaking triumphs, and a robust mission with the appropriate deterring capabilities fulfils it. The presence of a strong mission force with a proactive mandate to protect civilians deters action by groups that aim to subvert or sabotage the peace process or the mandate and secure an enabling atmosphere for diplomatic solutions. Robust peacekeeping, therefore, becomes a necessary element of conflict management and plays a major a role in peacemaking. UNAMSIL clearly portrays the contribution of robust mandates to creating an enabling environment for diplomatic resolution of conflicts. In the Sierra Leonean mission, Security Council resolution 1270, which established UNAMSIL, did not authorize UN forces to use force to counteract the attacks of the Revolutionary United Front (RUF) rebels on civilians. Consequently, UNAMSIL forces and the civilians suffered mounting attacks from the RUF rebels, which resulted in the killing of civilians and nine UN peacekeepers and over 500 peacekeepers were taken hostage.[75] Because of UNAMSIL's incapacity to protect Sierra Leoneans under the traditional peacekeeping mandate, the Security Council adopted resolution 1313, which authorized UNAMSIL "to deter and decisively counter the threat of RUF attack by responding robustly to any hostile action or threat of imminent and direct use of force."[76] UNAMSIL did so and achieved success in protecting civilians, which ultimately led to the resolution of the Sierra Leonean conflict and establishment of democracy in Sierra Leone.

UN peacekeeping missions are launched in violent civil wars replete with serious civilian killings and are expected to protect civilians, but they have failed because they operate in a "gray area" between traditional peacekeeping and peace enforcement with no clearly defined doctrine. The

United Nations seems to operate in a vaguely defined no-man's land, lying somewhere between traditional peacekeeping and peace enforcement, called the gray area, for which it lacks any guiding operational concept.[77] The United Nations finds itself in trouble in this gray area because it is wrongly applying perfect traditional tools to "inappropriate circumstances," violent civil war environments that demand a robust response from peacekeepers. Unfortunately, peacekeepers are hampered by strict adherence to noninterventionist, traditional principles of peacekeeping originally designed for interstate conflicts. There is, therefore, a need for a paradigm conceptually distinct both from impartial peacekeeping based on consent of the parties and enforcement action to bridge this doctrinal void. The basic thrust of robust peacekeeping is to fill the doctrinal gap by deploying deterrent forces in conflict zones with limited consent from parties and wider use of force than in traditional peacekeeping but stopping short of war. Enforcement, by contrast, is a military intervention, authorized under Chapter VII, for which Article 42 of the UN Charter permits United Nations forces to "take such action by air, sea, or land forces as may be necessary to maintain or restore international peace and security."[78] It is designed to impose public order by force with or without the host country's consent. An enforcement mission presupposes an identifiable aggressor who becomes the target of enforcement action, as in the 1990 Gulf War. These operations are usually beyond the United Nations' ability to command, control, and plan, and therefore are carried out by a coalition of countries or a regional organization such as NATO. The UN-sanctioned enforcement missions, some attempted under joint United Nations Command and led by a major power, often with disillusioning results. These took place in Korea (1950), Congo (1960), Iraq (1990), and Haiti (1994). Tables 3.1 a, b, c in Appendix 2 detail the differences in the mandates of traditional peacekeeping, robust peacekeeping, and enforcement actions in respect of the rules of engagement that determine the level of force to be used by forces.

Contrary to the rules of engagement in the traditional UNAMIR mission that proscribe the use of force to protect civilians under threat, the rules of engagement of the robust MONUC established in 2000 permit the use of necessary action to protect civilians under threat. Security Council resolution 1291 (2000) of February 24, 2000, established under Chapter VII of the United Nations Charter, authorizes MONUC forces to take "necessary action in their areas of deployment as it deems fit and within its capability, to protect themselves, collocated Joint Mission Contingents and civilians under imminent threat of physical violence."[79] Table 3.1 c shows the rules of engagement for the UN enforcement action in Iraq in 1990. United Nations Security Council resolution 678 (1990) authorized Members to use all necessary means to uphold and implement resolution 660 (1990) if Iraq did not implement the resolution by January 15, 1991.

The effectiveness of robust peacekeeping in ensuring civilian protection and stability is informed by the rational choice theory of civil war termination and demonstrated in the *formidable barrier* model of peacekeeping success model, to be discussed in detail later, which entails calculations parties make to support or reject peace. The theory presumes that the decision to fight or not is a function of the cost of continued fighting and the benefit of achieving a settlement. To achieve peace, the cost of war must be increased, but that can depend on UN forces' capacity to increase the cost of continuous fighting and the benefits of settlement for combatants. The underlying strategy of robust peacekeeping with a deterrent posture is to alter the parties' calculations of the cost of continuing to fight compared to the benefits of choosing peace. The basic presumption in using deterrent force to restore stability is that hostilities harden bargaining positions and attitudes rather than create concessions by parties who suffer costs. The probability of accepting a diplomatic resolution to a conflict by parties afflicted by human and other related costs in an ongoing hostile conflict depends on the cessation of hostilities and attacks on the groups' members. A deterrent force can create stability for political dialogue to resolve the conflict. Deterrence is, however, aimed not at favoring any of the parties, but rather at enforcing the mandate.

Dilemmas of robust peacekeeping

The paradigmatic transformation to robust peacekeeping has triggered political and operational dilemmas that threaten its potency to protect civilians because robust peacekeeping contradicts the constitutive principles of traditional peacekeeping doctrine, namely consent, impartiality, and limited use of force for self-defense. Challenges to robust peacekeeping include Member States' disagreement on the definition of the doctrine of robust peacekeeping, disagreement on the use of force, prospects of open battles with armed groups, a greater threat to nonmilitary UN personnel, reprisals against civilian populations, lack of consent and cooperation from conflict parties as well as extremists and spoilers. Lack of adequate resources, absence of major power participation, absence of universal operational training doctrine for UN troops, and the politics of civilian protection by the Security Council are other structural challenges likely to undermine the deployment and operational efficacy of robust missions to protect civilians when deployed in violent civil wars. I discuss these in detail in this section.

Major key UN players lack consensus on the doctrinal definition of robust peacekeeping and the actions that fall legitimately under it. The disagreement among UN members, particularly those on the Security Council, is centered on the appropriateness of the use of force in UN peace operations between states that uphold the principle of sovereignty and those that support humanitarian intervention. In spite of the emerging consensus that the United Nations should apply force as a last resort to protect civilians, many states such as China and Syria strongly uphold the principles of minimum use of force for self-defense and non-intervention in sovereign states, as China argues in the case of the Darfur and Syrian conflicts. These protectionist states conceptualize the use of force by UN peacekeepers as an intrusion on the sovereignty of the host country in contradiction to Article 2(7) of the UN Charter, which protects state sovereignty and domestic jurisdiction. Acting on this principle China and Russia frustrated the deployment of a robust mission in Darfur where a more deterrent force to protect civilians from violations was required.[80] The problem is that the UN Charter concurrently promotes the two fundamental values of universal human rights and sovereignty, which are diametrically opposed to each other, thus generating tensions between them. The UN Charter has several specific provisions that protect national sovereignty and, concomitantly, protect states from the use of force by other states. Article 2(4) of the UN Charter states, "All members shall refrain in their international relations from the threat or use of force against the territorial integrity or political independence of any state or in other manner inconsistent with the Purposes of the United Nations,"[81] and Article 2(7) of the UN Charter also states "Nothing contained in the present Charter shall authorize the United Nations to intervene in matters which are essentially within domestic jurisdiction of any state." However, the Charter outlines the exceptions to the protection of sovereignty in Article 42 of Chapter VII of the UN Charter, which grants the UN Security Council the power to authorize the use of force to intervene in a sovereign state after it has found under Article 39 that there exists a threat to peace, a breach of peace, or an act of aggression.[82] Since protection of human rights is a purpose of the United Nations and has been accepted by Member States to transcend state boundaries, the use of force by UN forces to protect human rights arguably falls under the exception clause of the use of force. Despite the United Nations' legitimacy to use force to protect civilians, anti-force UN members, particularly China, continue to frustrate the authorization and deployment of robust missions in violent conflicts, irrespective of the severity of human rights violations.

This tension has trickled down into the realm of UN peacekeeping operations. In the field of peacekeeping, the dilemma is premised on the fact that originally UN peacekeeping was doctrinally and operationally designed to prevent UN peacekeepers from getting embroiled in conflicts and tilting the balance of power on the ground in favor of a conflict party.[83] The interventionist's posture of robust peacekeeping, which conceptually demands UN peacekeepers to use force to deter or confront spoilers that attack civilians, has been criticized by noninterventionist key UN players such as Russia and China whose interests shape the Security Council's decision to deploy missions. This political divide influenced the debates over Darfur and the subsequent deployment of a weak, hybrid United Nations–African Union Mission in Darfur (UNAMID) with a watered-down robust mandate that incapacitated its ability to protect Darfurians from human rights violation. Opposition from noninterventionist civilian components of integrated UN peacekeeping missions may also frustrate the operational efficacy of deployed robust missions. Contemporary UN peacekeeping missions operations are mostly integrated ones whose effectiveness depends on successful coordination, communication, and cooperation among the relevant field actors. However, some civilian components of integrated missions such as UNDP, UN OCHA, and UNHCR, that are averse to the use of force and have taken clear stances against being linked to any military effort, often find it inappropriate and unconducive to be linked to UN missions whose robust rules of engagement permit the use force to fulfill their mandates. This ultimately undermines the operational success of the mission

A more serious dilemma facing interventionists is the absence of conceptual clarity on the appropriate level of force to be used by mission forces in this gray area between "peacekeeping" and "enforcement." The DPKO and the DFS's *Capstone Doctrine*, which aims to create a set of common guidelines for UN planners and practitioners, recognizes the need for a clearer articulation of the doctrinal foundations of UN peacekeeping operations but struggles to articulate clear guidelines on the use of force.[84] While the document upholds peace operations' adherence to the principle of nonuse of force except in self-defense and defense of mandate, it at the same time counterintuitively underscores the fact that peace operations can use force proactively to defend their mandate and protect civilians under threat. This contradictory position sends confused signals and presents incoherent guidelines to military commanders and field practitioners on the use of force, culminating in differential interpretation and level of force used either by UN forces in different missions or by different troop contingents within the same mission. The dilemma is how much minimum force is enough to deter spoilers, while at the same time avoiding tipping into the sphere of enforcement? The danger is that too little force could lead to mission failure to protect civilians while too much could be deemed excessive and elicit a violent backlash from spoilers along with global condemnation. In Rwanda and Srebrenica, the United Nations was criticized for being too restrictive in using force, resulting in the genocides whiles in Somalia some troops were accused of using excessive force that led to attacks and the killing of UN forces by certain Somali factions.

Moreover, wherever UN robust peacekeeping missions with protective mandates are deployed in civil wars fraught with attacks against civilians, they raise expectations among the local population and the international community that this will protect civilians against violations by spoilers, and that failure to live up to these expectations will undermine trust and confidence in the United Nations' ability to protect human rights and "to save succeeding generation from the scourge of war."[85] The use of force, or the failure to use force, then impact the confidence of the local civilian population and the broader international community in the United Nations, as well as in the safety of the missions' civilian and military partners in the areas of operations.[86] The dilemma is finding the right balance in applying force that avoids popular resistance and wins the trust of the people. The use of force also raises issues of the legitimacy of the United Nations among local populations. Excessive force, can be perceived as an invading or imperialistic force, or too heavy handed, eliciting reprisal from some conflict parties as was the case of UNISOM in Somalia. However, lesser force makes peacekeepers ineffective and lack of civilian protection affects a mission's legitimacy and credibility as a peace agent among the populace. Their perception of the United Nations' ability to protect them wanes, as in the case of UNAMIR, which was closed prematurely. In the eastern Democratic Republic of Congo (DRC), MONUC's failure on several occasions to use force to protect the civilian population from indiscriminate attacks led to violent protests and attacks against these forces. Such provocative attacks on UN forces undermine the missions' reputation for robustness. Since UN forces will not fire back against civilians, it gives the false impression that their firepower is essentially symbolic. The cases of Rwanda and the DRC illustrate this dilemma. There are indications that MONUC over time has experienced a loss of legitimacy among the civilian population in eastern DRC, partly because of its failed "robust" stance.[87]

The use of force in robust peacekeeping may seem to compromise a mission's impartiality and neutrality and that undermines their effectiveness, erodes confidence in them, makes them targets of attack, and elicits withdrawal of consent and noncooperation from afflicted conflict parties. In traditional peacekeeping consent of conflicting parties is a criterion for mission success, because it is necessary for trust building and peaceful negotiations. Missions deployed with conflict parties' consent are less threatening and more acceptable than nonconsensual missions to conflict parties, and also reduce the risk of combat casualties to mission troops. The absence of consent is a guarantee for noncooperation, as evidenced in Bosnia where Bosnian Serbs disregarded and attacked the safe area and in Somalia where UN forces suffered high rates of casualties. The lengthy negotiations between the United Nations and Khartoum that delayed the deployment of a hybrid UNAMID mission with a watered-down robust mandate to Darfur illustrates that, despite its increased focus on "robustness," the United Nations still needs consent of key conflict parties to be a credible, robust, effective intervener. The book disagrees with the argument that once conflict parties consent to the deployment of a mission, it would guarantee their cooperation with the mission. The correlation between consent and cooperation by conflict parties seems overblown. For example, the Khmer Rouge and the Phnom Penh regime consented to the deployment of UNTAC but they refused to cooperate and disarm. The principle of consent is no longer absolute because of the changed contemporary conflict environment, as well as the conflict parties. Peacekeeping missions in the Cold War era were mostly deployed in interstate conflicts with the consent of conflict parties after a peace agreement where there was usually a peace to keep. However, in post–Cold War conflicts, the United Nations cannot obtain the consent of all conflict parties, especially the predatory rebel groups that gain from continued conflict and have no interest in organized political structures. To be able to operate effectively in these volatile conflict environments at all, the United Nations has found it expedient to modify the principle of consent to some degree, so that missions are deployed with the consent of the main or relevant conflict parties. The problem is to determine the relevant conflict parties. Because of this choice, the United Nations, however, risks being accused of selectivity and establishing differential relations with the conflict parties. Such a situation has a dangerous security ramification of becoming a part of the conflict raising issues of its legitimacy as an honest and impartial broker of peace as in the case of MONUC in the DRC where MONUC obtained the consent of the government but did not receive the consent of several important parties.

Neutrality and impartiality are principles underlying traditional UN peace missions designed to facilitate intervention and forge a solution that meets the interest of conflicting parties. Robust peacekeeping, which takes a more activist approach, deemphasizes impartiality in an effort to protect civilians. However, using force challenges the United Nations' recognition as an impartial player and as a nonpartisan peace broker. However, egregious civilian killings in contemporary civil conflicts necessitate a redefinition of the principle that differs from its traditional conception. Impartiality means that UN forces should not be prejudicial to the claims and positions of the conflict parties, but the UN forces can be partial in enforcing a peace agreement or mandate and in using force against a spoiler. Otherwise the mission risks failure to protect civilians as well as being accused of condoning the spoilers' behavior censored by the international community. Peacekeepers can be prejudicial or partial when it comes to mandate enforcement, including civilian protection. Historically, the United Nations has not always upheld the principles of impartiality and nonuse of force. ONUC forces in the Congo (former Zaire now DRC) in the 1960s were compelled to use offensive force to neutralize some of the "spoilers," or secessionists, in the Katanga province. The basic argument is that impartiality is not a hindrance to civilian protection. In robust peacekeeping, therefore, peacekeepers are impartial in fulfilling the mandate. Impartiality means that they require all parties to the conflict to comply with the peace agreements including certain minimal standards of human rights. According to Under-Secretary General for Peacekeeping Jean-Marie Guehenno, "The United Nations maintains its impartiality toward signatories of peace agreements but we are definitely partial towards implementation of the agreement. We are not neutral when the agreement is under attack."[88] The mandate to protect civilians in robust peacekeeping missions seems to discourage strict adherence to the principles of impartiality. The United Nations' partial implementation of the mandate using force bolsters its credibility in the eyes of the local populace and the international community, contrary to traditional peacekeeping. However, some conflict parties continue to accuse UN forces of partiality, as in the case of MONUC, which carries out joint operations with the Congolese army against the rebels in eastern DRC.

The use of force, it can be argued, makes mission forces vulnerable to attack and increases reprisals against civilians by combatants, as evidenced in attacks on UNISOM II and UNPROFOR missions' forces. In Bosnia the use of force by NATO led to the capture of UN forces and attacks against civilians. According to Sharp, deterrence is assured less by weaponry and more by civilian-based, nonviolent actions.[89] However, not using force to protect civilians does not guarantee the safety of either civilians or mission forces. In missions such as United Nation's Assisted Mission in Rwanda (UNAMIR), where the Security Council did not authorize them to use force to protect civilians, we witnessed genocide. Britain and India opposed a tougher mandate that used force for the Sierra Leonean mission because they feared exposing their troops in Sierra Leone to attacks from the RUF. That did not prevent the capture of British soldiers and reprisals against the civilians. Britain had to use force to free its troops held hostage by the RUF. UNAMSIL deployed in Sierra Leone was initially restricted to the use of force for defense only, but the predatory RUF rebel forces, whose prime focus was to loot Sierra Leone's diamond, continued attacks on UNAMSIL forces and vulnerable civilians. UNAMSIL was able to protect itself and civilians when the UN Security Council later handed it a robust mission with robust rules of engagement using force to protect civilians from the RUF's attack.

The reality is that, irrespective of restrictions on use of force, the *modus operandi* of spoilers in contemporary volatile civil wars is to target interveners and civilians to achieve their objectives. In a crisis situation that demands the stoppage of atrocities such as a genocide, the deployment of nonlethal peacekeeping forces on the posture of self-defense clearly becomes inapplicable and inhumane as manifested in the tragedy of the Rwandan genocide and the Bosnian massacres.

Another dilemma is that the phrase "all necessary means" in relation to the use of force in robust mandates is vague and ambiguous, leaving the decision to use force to the missions' leaders and field commanders of the contributing contingents. They are in most cases guided by their domestic legal systems to determine the use of force. We witness differential interpretations and possible conflict between United Nations' rules of engagement and that of contributing nations in respect to the use of force. The United Nations' rules of engagement authorizing forces to use "all necessary means" including force to restore peace and security lose its resonance when UN troops are constrained by their national laws.[90] Mission contingents' reliance on their domestic legal systems to guide the use of force emanates from the participatory precondition that contributing states have disciplinary jurisdiction over its troops, and not the United Nations. As a result, participating contingents, even in robust forces, are affected by the "phone home" syndrome in which participating contingents seek domestic legal guidance in respect to the use of force. Clashes between the missions' national contingents' approaches and doctrines and interpretations on the use of force within a mission can lead to incoherent and ineffective peace operations. The differential application of force within the same mission would also raise questions about the missions' legitimacy. Without common, or at least congruent, understandings about basic principles of rules of engagement, decisions made wither in a conceptual void,[91] culminating in operational or doctrinal incoherence and mission failure. Inconsistency in the rules of engagement can both impede streamlining command and control, and also increasingly complicate efforts to form unified operational doctrine.[92] Some assume that once a national contingent is incorporated into a UN peacekeeping force and placed under United Nations' command, the United Nations has supranational authority over the contingents,[93] but that assertion relies on the participation agreement for national contingents, which only deals with provisions concerning jurisdiction, logistics, and compensation.[94] Rowe highlights this conflict in his analysis of the British experience in Bosnia and Herzegovina where British soldiers participating in the United Nations Protection Force (UNPROFOR) may commit an offense under British law by not using force in certain situations under the national rules of engagement system, but may be violating the UN rules of engagement if they use force. Rowe asserts that under British law UN rules of engagement have no binding status, and thus a British soldier's actions are entirely assessed in accordance with British criminal law.[95]

The complex and divisive politics within the Security Council fuelled by neorealist national interest considerations, also poses a hindrance to the deployment of robust missions. Security Council members' national interests have impacted the designation, deployment, and provision of resources for missions. Divisive power struggles among the five permanent members of the Security Council in most cases result in vague, ambiguous, or compromised resolutions. These are designed to satisfy members' interests to avoid gridlock and obtain the consent of all Council members to enable the deployment of a mission, irrespective of its capacity to respond to the existing and potential threats to civilians. These political compromises and consequentially restricted rules of engagement in most cases undermine the potency of mission forces to apply appropriate force to protect civilians from imminent human rights violations by spoilers. The Rwandan genocide and Darfur killings are clear manifestations of the failures of political will and moral obligations due to divisive politics within the Security Council fueled by neorealist considerations of prioritizing national interests over global collective interest. Despite international and conflict parties' support for a robust mission in Rwanda as articulated in the Arusha peace accord, the United States' disinterest in the conflict as a result of the Mogadishu debacle where 18 Americans troops were killed, resulting in a poorly designed traditional mission that could neither protect itself nor protect the defenseless civilians massacred. In the case of Darfur, China's and Russia's opposition, due to their neorealist economic, military, and political considerations and ties with Khartoum government of Al Bashir, resulted in the deployment of a hybrid UNAMID mission with a watered-down robust mandate that rendered it incompetent to protect civilians from atrocities.

A considerable obstacle to a successful robust operation is the major powers' disinterest in contributing troops to peacekeeping missions, especially in conflicts with no direct effect on or of less strategic interest to them. Major powers' availability for UN peacekeeping missions is contingent on their national interests and their calculation about risk to troops and sustainability of domestic support. Many states with advanced military capabilities contribute fewer troops to UN forces, or are deeply selective in deploying troops in conflicts. Today, fewer than 2 percent of UN troops deployed in African missions come from Europe or North America.[96] Traditional peacekeeping was designed to avoid major power participation due to concerns about Cold War competition, but the new paradigm of robustness emphasizes major power participation because they have the capacity for mission success. Nevertheless, this seems problematic for a variety of reasons, including perhaps left-over suspicion of peacekeeping and bitter experiences with it, so they are reluctant to contribute troops. Unlike peacekeeping for middle or small power armies, major powers reportedly have seen such duty as ancillary to the main mission of the armed forces, and in some cases damaging to military promotion and careers based on combat duty. In addition, several of the world's most capable militaries, including the United States and Britain, are reluctant to contribute troops to UN missions because of their involvement in other commitments deemed pertinent to their national security interests. They have focused on the priority of the fight against global terrorism, while others such as Russia, China, and Japan contribute few personnel for doctrinal reasons. The United Nations today relies on a comparative handful of 15 states, particularly South Asian and African states, that regularly contribute about 75 percent of UN forces.

The casualty tolerance level of a contributing state and calculation about risk to troops and sustainability of domestic support also affect their availability for robust missions. As Sir Jackson argues, there is no such thing as risk-free soldiering.[97] Risk-averse states with low casualty tolerance levels and those, especially major powers, with limited strategic interests in the conflict, do not contribute troops. There is a general lack of willingness to accept casualties while fighting nonexistential threats. Risk-averse states that contribute troops strive to minimize the exposure of their troops to danger and avoid deployment in dangerous parts of the mission theater. Closely related to the cost or risk tolerance explanation is public support for the deployment of nationals in conflicts that do not directly impact on their interests. The more democratic the state, the less inclined it may be toward such risk, given the potential for domestic political backlash should national personnel be lost. Wealthy states also stress loss avoidance, and most of their commitments to UN operations are monetary in nature, or providing logistical support, sometimes through private sector contractors.[98] Contemporary public opinion in many Western liberal democracies is that current threats from civil wars are perceived as less imminent or existential than they did a few decades ago with general wars.[99] The majority of the populations doubt the effectiveness and legitimacy of using force to promote peace and stability, especially in "unnecessary wars" and sometimes view the use of force as a potential trigger of new threats. Military involvement in "unnecessary wars" that do not directly threaten their interests therefore becomes harder to sell to the populace of Western democracies than the overhanging threat of nuclear annihilation and terrorism.[100] Their aversion to "unnecessary wars" has been compounded by the contemporary economic crisis and cost of involvement in wars. The United States' leading role in NATO's forceful intervention in Libya, which established the "no fly zone" over Libya and helped the rebels overthrow the Gaddafi regime, was hotly debated domestically by the public despite the clear picture of ongoing crimes against humanity and war crimes. The Obama administration had to lead the operation from behind in order to garner public support and avoid a political backlash. The levels of massacre were not enough to command popular support for the use of force in Libya as in the case of Rwanda. Democratic governments, fearing electoral backslash for ignoring public opinion, must balance between their loyalty to international organizations and their need to maintain domestic support for their national agendas.

The United Nations acknowledges the problem of resources, particularly funding, personnel, logistics, and equipment, since members have been reluctant to honor their financial obligations. For example, as of February 28, 2005, Member States owed approximately US$2.25 billion in dues. The largest debtors were the United States (US$743 million) and the Russian Federation (over US$500 million). As a result, there is a growing disparity between the capacity of the United Nations and the demands for peace and security that require resources. Secretary-General Annan, in a report on the role of the United Nations in the twenty-first century, highlights some of the limitations of the United Nations by comparing the UN system of launching operations to a volunteer fire department, because "every time there is fire, we must first find fire engines and the funds to run them before we can start dousing any flames."[101] General Romeo Dallaire, Force Commander of UNAMIR (Rwanda), also sees the United Nations as a "pull" and not a "push" system. In a "push" system such as NATO, basic resources such as food, fuel, and ammunition, needed by personnel, will automatically be provided. In a "pull" system, such as the United Nations, you have to apply for everything, regardless of size and type of mission.[102] Since 1999, the United Nations has authorized robust Chapter VII missions but the missions suffer from a severe mandate–resource mismatch. UNAMID and MONUC are examples of strenuous limited resources and how missions are unable to keep massive civilian–military endeavors operational in harsh environments. These missions are bereft of adequate military personnel, equipment, and logistics to implement complex mandates, to protect UN forces and the civilian population in the conflict areas. Corrective measures proposed include global tax on currency taxes, international transportation, and arms trade, but major powers are reluctant to make reforms, fearing a loss of political leverage.[103] The UN logistics base in Brindsi, Italy, has, however, been stocked with a "strategic equipment reserve" with enough weaponry and equipment to deploy within 30 days for smaller missions and 90 days for larger missions.[104] Despite these efforts, current robust missions do not have enough resources to execute their mandates including civilian protection. The United Nations now finds itself at a tipping point, where robust ambitions clearly outweigh actual capacity for robust action: resources, absence of major power participation, quick deployment. UN peace operations seem to be heading toward a state of "conceptual overstretch" defined as a perceived gap between increasingly ambitious UN concepts and the organization's lack of political and military capacity to operationalize and implement these concepts.[105] MONUC's incapacity to protect civilians from attacks by spoilers in eastern DRC vividly confirms the wide gap between UN rhetoric and the actual will and capacity to be robust on the ground.

More disturbing is the lack of a common training doctrine for mission personnel, which undermines mission coherence, coordination, command, and control. The *Capstone Doctrine* does not "override the national military doctrines of individual members states,"[106] depriving the United Nations of the unique opportunity for a universal functional military training doctrine to guide preparation of troops for deployment. Compared to NATO, the UNDPKO and the DFS suffer from a severe lack of military planning capacity,[107] which ultimately dampens the political will of Western nations to commit troops to UN peace operations. A robust mission in a dangerous and complex conflict environment would require a coherent and streamlined UN training doctrine similar to NATO's to guide the mission. The UN training doctrine should override national training doctrines since such diverse doctrines would ultimately result in policy and operational inconsistences, and incoherence that would incapacitate a mission.

Despite these drawbacks, robust peacekeeping missions are increasingly demanded and mandated because, unlike the noninterventionist traditional peacekeeping, it constitutes a formidable barrier to human rights violations in civil wars. In this study, I borrow from the rational choice theory of conflict termination to argue that mission success would be attained if the mission has the capability to alter the calculations of conflict parties from continuing to fight to choosing peaceful negotiations. Following the logic of the costs of war termination theories, I develop the *formidable barrier* model of UN peacekeeping mission success. The rational choice theory of war termination and the *formidable barrier* model of peacekeeping success are discussed in the next two sections.

Rational Choice Theory of Conflict Termination

The rational choice theory of civil conflict termination postulates that the decision to fight or not is a function of the cost of continued fighting and the benefit of achieving a settlement.[108] War will occur if combatants make a calculation that the benefits of escalating a conflict to a violent confrontation outweigh the costs, while peace will take place if they perceive that the continuance of war threatens their survival and political objectives more than the abandonment of hostilities. Consequently, the combatants' decisions on their strategies of war and peace will be determined by a rational calculation of the constraints and opportunities that confront them at any given time. The constraints and opportunities are basically a function of each combating party's ability to manipulate the relative costs and benefits of continued fighting. The *expected utility theory*, a variant of rational choice theory, postulates that the decision to fight or negotiate is determined by the relative costs of and benefits of a unilateral victory or a compromise settlement.[109] The theory postulates that war is a rational process consisting of careful deliberate calculations by leaders who are rational expected utility maximizers. Combatants carefully estimate their chances of winning a civil war, the amount of time they will take to achieve this victory, how much it will cost, and the relative payoffs from winning versus accepting a settlement. Settlement occurs when combatants believe they can do no better by continuing to fight than by bargaining. The risk of war takes place within a framework of circumstances that comprise a set of necessary conditions that must be satisfied before war is initiated. The choice of war therefore depends on estimated costs and benefits. The decision to go to war is guided by a desire to maximize the net benefit expected of it. The group's welfare is intimately tied to the overall costs and benefits imposed on the group by the choices. The group will not start a war or continue to fight if it perceives the net expected result to be less beneficial than that of restraining war, remaining at peace, or surrendering to the opponent

The classical *power preponderance theory* of war termination also postulates that military power superiority of a conflict party or asymmetrical distribution of military power among conflict parties ends hostilities and fosters negotiated peaceful settlement of disputes because it invites compromises from the weaker party.[110] Power preponderance destroys the possibility of a guaranteed and easy victory and alters the calculations of the weaker conflict party who will neither risk initiating conflict nor continue with hostilities in view of the formidable military power of the stronger opponent. According to this argument, the risk of war between two parties is lowest when a party to the conflict is overwhelmingly stronger than the other. Advocates of "power preponderance" argue that "parity is the necessary condition for major war."[111] Power parity between two conflicting parties creates uncertainties about each other's military capabilities, sometimes leading to underestimation or miscalculations of the opponent's capability, which increases the risk of war in a realist competitive environment. War occurs when at least one of them is uncertain about the capabilities of the other. This uncertainty is negligible and a peaceful negotiated settlement of the conflict is reached when a conflict party has a preponderance of power than the other.[112] The greater the challenger's uncertainty and estimate of the distribution of power, the closer the parties are to parity, and the higher the risk of conflict between the two.[113] The *power preponderance* argument is closely related to Organski's *power transition* argument that major power wars occur when the challenger obtains similar capabilities as the hegemon and attempts to overtake it. Closely related is the *hegemonic stability theory* that postulates that a stable international system is guaranteed when a single state with a preponderance of power dominates the international system.[114] A diminished power of the hegemon triggers violent response from the rising challenger, which destabilizes the international system. Contrary to the *power preponderance* argument, advocates of the *balance-of-power* theory claim that "equality of power destroys the possibility of a guaranteed and easy victory and therefore no country will risk initiating conflict."[115] Wars occur when there is power preponderance, "because there is nothing to prevent them."[116] According to this argument, where there is power parity, the risk of war between two states is the lowest. Recent empirical studies of the determinants of war, however, tend to support the "power preponderance" argument.[117] Following the logic from the costs of war termination theories, I have developed the *formidable barrier* model of UN robust peacekeeping mission success. I test them in chapters 5 and 6.

Formidable Barrier Model of UN Robust Peacekeeping Success

Based on the cost–benefit theories about decisions on war and peace in civil conflicts and of civil conflict termination, I have generated the *formidable barrier* model of UN robust peacekeeping success for staving off deliberate civilian killings by combatants in civil conflicts. This model posits that the success of UN peacekeeping missions in staving off deliberate civilian killings is a function of its deterrent capacity. In situations of continuing and recurrent violence, sometimes described as enduring rivalries even within states, a more formidable and less passive barrier may be required, and peacekeeping missions must be more robust and extremely impressive in strength or excellence with the capacity to impede the intentional killing of civilians by combatants in a civil conflict. In this model, deliberate civilian killing, as defined and measured by Kreutz et al. (2005) (Uppsala Conflict Data Program one-sided dataset), is considered as an instrument for advancing political goals such as territorial control, sometimes now referred to as "cleansing."

Borrowing from the cost–benefit theories, the *formidable barrier* model posits that war as well as deliberate civilian killings is a rational process consisting of careful deliberate calculations by rational actors. As a policy instrument, intentional civilian killings should subordinate to the political calculations for achieving combatants' political goals. The political calculations make it possible for combatants to perceive civilian killings and peace as alternative policies and to tradeoff between them. The choice of continued fighting depends on the estimation of costs and benefits to the leaders or combat group. Combatants will not continue to fight and kill civilians if they perceive the net expected benefits to be less than that of peaceful negotiations. A group's probability of success or loss, perception of benefits and costs of continued fighting is a function of its relative capabilities compared to that of the intervener in this case robust UN mission forces. To initiate war, therefore, the spoilers must consider the relative strength of the mission force. A group's calculation of expected gains and losses of war also depends on the relative capability advantage the attacker has over the mission force. A group will continue to fight and kill civilians when it believes it is stronger than the mission force.

Peacekeeping missions' success in staving off deliberate civilian killings therefore is a direct function of their ability to constrain combatants' violent behavior. This results in a cost–benefit calculation by the combatants that lead to the choice of peace rather than continued fighting and civilian killings. In this sense, the mission's capacity to raise the cost of continued fighting by combatants is dependent on its relative capabilities compared with that of the combatants. A stronger mission force that has a large number of troops and stronger firepower reflected in its robust rules of engagement and stronger combative and repulsive capacities may alter the calculations of combatants and spoilers from continuing fighting and killing civilians to choosing peace. In the face of the overwhelmingly robust UN mission forces' preponderance power, rational combatants will opt for peaceful negotiations rather than a confrontation with the mission force to avoid incurring heavy losses. In addition to the physical strength, both the UN forces' and spoilers' cost-tolerance or willingness to suffer losses is of considerable importance in determining the formidability of mission forces to halt spoilers' atrocities. For example, in Somalia, despite their overwhelming physical strength, the UN forces had a low cost-tolerance and unwillingness to suffer casualties, which rendered it ineffective in the face of the rebels' high cost-tolerance and willingness to suffer. Therefore, in addition to its physical capabilities, the mission force must also possess the willingness to fight as well as a high cost-tolerance level to be able to constitute a formidable barrier to atrocities.

Robust peacekeeping, by designation, has the ability to manipulate the relative costs of continued fighting, and civilian killings, and the benefits of peace. Robust peacekeeping missions, which are characterized by large force strength, use of heavy weapons, major power participation, the robust rules of engagement to use force for self-defense, and with a high cost-tolerance level, can provide the necessary barrier or impediment to intentional civilian killings by combatants and spoilers by imposing some military constraints on them. Large force size can effectively monitor large land areas as well as protect a greater number of civilians. Large force strength also increases the mission's capacity and capability to match spoilers in strength and ward off their attacks. Thus, a large strength can alter the strength ratio between mission forces and spoilers, making the spoilers less able to challenge the mission force. The use of heavy weapons by a robust mission force also increases its firepower and offensive and deterrent capabilities, which ultimately will alter the calculations of potential attackers to choosing peaceful negotiations in view of the likely costs of challenging a mission force with such overwhelming capabilities. The inclusion of major powers with well-trained forces and sophisticated equipment in the mission force enhances the mission force's firepower and combative capabilities. Major powers have the critical resources, both human and material, that a credible deterrent UN peacekeeping force needs. Also, the realization that the world community, including superpowers, supports an intervention gives the mission a stamp of seriousness and may exert pressure on protagonists to halt hostilities. Rules of engagement that permit use of force signal to combatants that violence would be an unacceptable option and would be reciprocated using formidable power. The assurance of reciprocal attacks from mission forces would increase the adversary's cost in terms of casualties and reputation. Thus robust rules of engagement in the face of increasing costs of attacks would deter possible adversarial spoilers. The basic argument in this model is that the costs of killing civilians are low for spoilers relative to peace in the absence of a formidable barrier. Peacekeeping has to be sufficiently strong and enough of a deterrent to make the cost of attacking and killing civilians higher. The model shows that the success of a peacekeeping mission in staving off deliberate civilian killings in civil conflicts is a function of its deterrent capacity. In addition to its physical capacity, the mission force must also have the willingness to fight in order to constitute a formidable barrier to atrocities.

Model 1 below shows a simple sequential game-theoretic model of the *formidable barrier* model of UN robust peacekeeping success described above. A strategic game is a model of strategic interaction among decision-makers or players in which the choices they make affect the other. Each player has a set of possible actions and their choices result in conceivable outcomes. Each player has preferences and receives payoff for the choices made. Each player is given a complete numerical scale with which to compare all logical outcomes of the game. The payoff for each player is the number associated with each possible outcome. The numbers are only educated guesses. Generally for each player, a higher number means a better outcome.

The model demonstrates the interaction between the spoilers and robust UN forces deployed in a civil war and rests on five assumptions. The first is that spoilers and UN forces are rational actors whose decisions are informed by a cost–benefit analysis. They calculate future consequences of their actions. Thus actions taken are to maximize their benefits at the least cost. They would therefore avoid an action that would incur them a comparatively a higher cost than the benefit. Second, I assume that both spoilers and UN forces know that their rival, who they confront over civilians safety, will act optimistically to maximize their benefits if given a chance. They do not think their rival will forgo the opportunity to maximize their benefit if given a chance (figure 3.1).

Third, the safety of the civilian population is guaranteed when they are defended by a powerful robust UN force against spoilers. Fourth, spoilers would be deterred from attacking vulnerable civilians with the deployment of a robust UN mission force with a preponderance of power relative to the spoilers' capabilities and with a robust mandate to defend civilians. Finally, robust missions, for the sake of its own and the United Nations' credibility and legitimacy as well as mission success, will defend civilians when attacked by spoilers.

In this model, the respective choices, decisions, and actions of both spoilers and UN forces affect the safety of vulnerable civilians in civil wars. Their decisions or choices are based on their respective expectations of the opponent's potential response. Spoilers have the choice to either attack civilians or not. UN mission forces have the choice to either defend civilians or not when attacked by spoilers. In this game, the spoilers' payoff is listed first and the UN forces' payoff is second. The game begins with a decision by the spoiler to attack or not attack

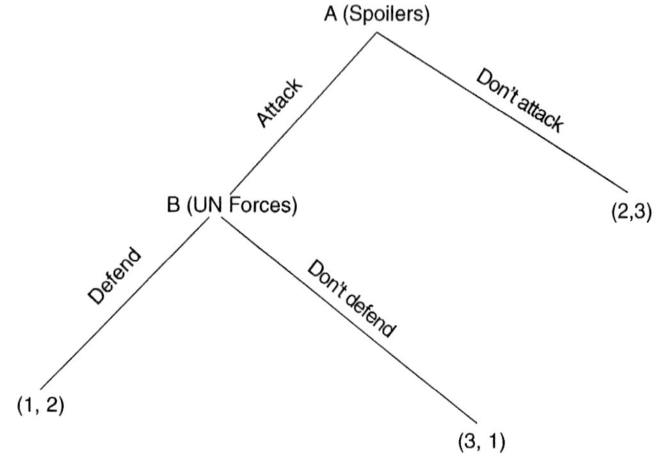

Figure 3.1 Game theoretic model of the *Formidable Barrier Model* of UN robust peacekeeping success

civilians. If the spoiler does not attack civilians, then there will be no confrontation with UN forces, creating the much needed political space for peaceful negotiation of the conflict. However, if the spoiler decides to attack civilians, UN forces have the option to either defend or not defend civilians.

The spoilers' set of actions are (*Don't Attack, Attack*) and UN forces sets of actions are (*Don't Defend, Defend*). Spoilers' preferences ordering, from best to worst, are (*Attack, Don't Defend*) (spoiler attacks civilians and UN forces do not defend them, so spoiler's payoff is 3 and UN forces' payoff is 1); (*Don't Attack, No UN Action*) (spoiler does not attack and no confrontation with the UN forces: Spoiler gets 2 and the UN forces get 3); *(Attack, Defend)* (spoiler attacks and the UN forces defend civilians, spoiler gets 1 and the UN forces get 2). Spoilers preferences are functionally represented as *(Attack, Don't Defend)* = 3 > (*Don't Attack, No Action*) = 2 > (*Attack, Defend*) = 1.

Robust UN mission forces' preferences ordering, from best to worst, is (*Don't Attack, No Action*) (spoiler does not attack civilians and avoid confrontation with UN forces, so spoiler gets 2 and UN forces get 3); (*Attack, Defend*) (spoiler attacks civilians and UN forces defend, so spoiler gets 1 and UN forces get 2); (*Attack, Don't Defend*), (spoiler attacks and UN forces do not defend, so spoiler gets 3 and UN forces get 1). UN mission forces' preferences are functionally represented as (*Don't Attack, No Action*) = 3 > (*Attack, Defend*) = 2 > (*Attack, Don't Defend*) = 1. Based on the preferences of both players, two sub-game equilibrium strategies are possible: (*Don't Attack, No Action*) and (*Attack, Defend*).

The most preferred outcome for the United Nations is the status quo where spoilers do not attack civilians and the worst outcome is when the UN forces stand by and do not defend vulnerable civilians being killed by spoilers, as happened in the Rwandan genocide. The most preferred outcome for spoilers is attacking civilians without protection or opposition from UN forces. Its worst outcome is attacking civilians who are defended or protected by the UN forces. In a real-world situation, most spoilers in most contemporary civil wars attack civilians to achieve their strategic goals and the UN mission forces deployed in civil wars today are mostly Chapter VII robust missions with robust mandates to use force to defend vulnerable civilians under imminent attack, in line with the contemporary humanitarian posture of the United Nations' peace operational paradigm. For the sake of mission success, the United Nations' legitimacy and credibility, and global confidence in its ability to honor its promise to protect human rights and "save the next generation from the scourge of war" as articulated in the preamble of its Charter, UN mission forces have to bear both legal and moral responsibilities to protect civilians. This can be achieved only when UN missions deployed have enough preponderance or deterrent power to constitute *formidable barriers* to the spoilers' quest to kill civilians, as in the case of UNAMSIL in Sierra Leone. A robust UN mission force with preponderance or deterrent power relative to the spoilers' capabilities coupled with the mission forces' willingness to accept risks to protect civilians from attacks can alter spoilers' calculations from continuing fighting and attacking civilians to choosing a negotiated peace. From the *formidable barrier* model of UN robust peacekeeping success, I articulate a set of hypotheses below and test them in Chapter 5 using UN peacekeeping data in civil conflicts from 1956 to 2006.

Hypotheses

From the literature on UN peacekeeping, I focus on the operational variables of peacekeeping relating to robust peacekeeping controlling for other possibly relevant factors in explaining mission success, such as structural and cultural variables to address these questions. In this study, the core operational variables to be used are mission type, mission strength, weapon type, resources (funds), rules of engagement (use of force), major power participation controlling for ethnic similarity, force makeup (cultural diversity or similarity), number of borders, conflict type, presence of peace agreement, and regime type. In line with this, I specify the following testable hypotheses and state the logic underlying them.

H1: Robust peacekeeping missions will be more successful than traditional ones in situations where parties are not yet in stable ceasefire or where rejectionists have organized spoiler opposition to agreements.

The basic argument is that a well-resourced robust mission with robust rules of engagement can deter spoilers from attacking civilians.

H1a: The larger/stronger the peacekeeping force, the more likely it will be successful.

H1b. The more resources (funds) a mission possesses, the more likely the success.

Fortna (2000) and Cox (1993) argue that the principal strategy of peacekeeping to change the behaviors of combatants is to raise the costs of continued fighting. One factor that increases the cost of fighting is a well-resourced, strong peacekeeping force. A well-resourced, large force can effectively monitor a large area and pose a formidable deterrence to spoilers.

H1c: Peacekeeping missions are more likely to succeed if troops are equipped with heavy, rather than small, weapons.

This variable is indicative of whether the force, in reference to the rules of engagement, is offensive (heavy weapons such as tanks) or defensive (small arms such as rifles). Type of arms also affects the deterrence capacity of the force. Heavy weapons have a better capacity to deter violators than small arms.

H1d: Peacekeeping missions are more likely to succeed if the rules of engagement permit them to use force to protect civilians from human rights violations.

The basic argument is that a mandate that prescribes use of force other than for self-defense only is necessary for peacekeepers to act and protect civilians threatened by spoilers. Spoilers are also discouraged and deterred from use of violence when they know that UN forces are mandated to use force to counteract threats to the mission objectives and protect civilians.

H1e: Peacekeeping missions are more likely to succeed when major power(s) participate in the peacekeeping operation.

The basic logic is that major power participation reflects the legitimacy and seriousness of global concern for the conflict. Troops of major powers are generally better equipped and better trained than those of minor powers. Major powers also have the political and economic leverages to influence the behavior of combatants by manipulating and raising the costs of continued fighting.[118]

H2: Decision-making calculations in assessing the cost of war and benefits of negotiations and settlements are higher in situations of robust peacekeeping than in traditional peacekeeping.

Decision-making is perceptual and, due to lack of data and measurement difficulties, I will test it in the case studies. I will look for indicators of cost–benefit calculations in the case studies.

Control Variables

In addition to the hypotheses described above, there are other variables that may affect civilian killings in a conflict zone that need to be controlled for. These are regime type, conflict type, peace agreement, borders, ethnic affinity, and troop composition.

H3: The greater the number of countries contributing to the mission contingents, the less likely the mission success.

Scholars have argued and supported with case studies that language and cultural differences among members of peacekeeping forces may have negative consequences.[119] Cultural and linguistic differences may result in disagreement in interpretation of a resolution and what actions to take. It is presumed that each country is culturally distinct.

H4: Peacekeeping missions are more likely to succeed if the highest field executive plus troop contingents have cultural or ethnic affinity with the host country.

One perspective is that peacekeeping forces are most likely to succeed if they are from the same region as the conflict. The argument is that affinity creates trust and legitimacy.[120] An opposite perspective is that regional peacekeepers are less likely to be regarded as neutral and trustworthy.[121] I use as a proxy the ethnicity of the highest field executive in determining whether the group affiliates ethnically or culturally with the parties to the conflict. The reason is that the tenor of the mission can be heavily influenced by the character and ability of the leadership.

H5: A mission is less likely to be successful in identity conflicts than in ideological conflicts.

Many studies on conflicts have suggested that identity conflicts (ethnic or religious) are far more difficult to resolve than ideological conflicts.[122] The reason is that identity conflicts are based on deep-rooted emotional values that are difficult to compromise on. On the basis of this realization, it presupposes that casualties will be more difficult to control in identity conflicts than in ideological conflicts.

H6: Countries with more bordering states see less peacekeeping success.

Neighboring states usually have a high stake in the outcome of conflicts and consequently act to either support or disrupt peaceful solutions. Some studies argue that neighboring states have the potential to disrupt a peacekeeping process with direct acts of violence or support to warring parties that oppose the operation (Diehl, 1994; Green et al., 1998). It is, therefore, anticipated that more borders will negatively affect peacekeeping success.

H7: Missions are more likely to be successful when deployed after peace agreements among conflict parties.

The argument is that the probability of peacekeeping success will be higher when there is a negotiated settlement to the conflict before peacekeepers are deployed than in the absence of such an agreement. Signing a peace agreement reflects the political will of the combatants to end the violent phase of the conflict.[123] Fortna (2003) argues that agreements employ several instruments to change the payoffs and make it costly to cheat, which reduces the uncertainty about compliance and intentions, and controls accidents.

H8: Mission success is likely to be greater when host states are democracies.

A major finding in the international relations literature is that democracies tend not to fight one another. This democratic peace proposition has more recently been extended to civil wars,[124] human rights,[125] and state-sponsored mass-murder.[126] Rummel (1994) finds that democracies have killed substantially fewer of their own citizens than other forms of governments.

Conclusion

In this chapter, I have examined the theories of robust peacekeeping as well as controversies about the likely outcome of robust peacekeeping missions. The chapter also presents a *formidable barrier* model of robust peacekeeping success derived from the rational choice theory of conflict termination. On the basis of these premises, I articulate the hypotheses to be tested in chapters 5 and 6. In the next chapter, I detail the methodology to test these hypotheses.

4

Data, Variables, and Methodology

This chapter details the variables, measurement, and data sources as well as the methodology. The thesis of this book is that robust peacekeeping works better than traditional peacekeeping in protecting civilians from deliberate killings by state and organized nonstate actors in civil wars. In the process we will also determine the effects on stated mission goals.

More specifically, as discussed in the previous chapter, the study seeks to answer the following questions: what accounts for UN peacekeeping forces inability to protect civilians from human rights violations, especially intentional killings? Is traditional peace operational doctrine, initially designed for interstate conflict, appropriate for the new intrastate conflict environment? Do mission characteristics of UN peacekeeping mandates and operations matter for reducing the ongoing intentional civilian killing? Does robust peacekeeping work better than other mission types in staving off civilian killings when UN peacekeeping missions are deployed in civil wars? To what extent does robust peacekeeping affect the number of intentional civilian killings? Is there such a detectable difference regarding other stated peacekeeping mission goals on a case by case basis? To answer these questions on UN peacekeeping, two approaches can be used. First, the comparative case study approach, on which some scholars rely.[1] The problem with using only case study approach is that it reflects small *N* and circumstances in those studies and therefore does not allow broad generalization across the population of cases excluded from the study. Some scholars argue that the use of case study alone to draw general conclusions is flawed by the problems of selection bias, lack of systematic procedures, and inattention to rival explanations.[2] Other scholars alternatively argue that these critiques of qualitative case research may be overdrawn and that contributions of these works are underappreciated.[3] The second approach is Large *N* quantitative aggregate data analysis. Certain scholars also use strictly quantitative analysis to generalize or make causal inference.[4] However, this methodology may lack the nuances that qualitative methods reveal.

The debate on the appropriate mode of analysis has illuminated both the usefulness and shortcomings of each methodological approach. Both King et al.'s (2004) *Designing Social Inquiry* and Collier's (2004) *Rethinking Social Inquiry* have demonstrated the usefulness of both methodological approaches in achieving social scientific ends but also acknowledged the shortcomings of each as a research tool. In view of this methodological divide, other scholars[5] proposed *integration* of the methodological approaches referred to as the *mixed method*, which Lieberman (2005) calls the *nested analysis*.[6] This unified approach combines case-study analysis with statistical analysis, thus, offering the advantages of each approach and also making up for the shortcomings of each. For example, statistical analysis can guide case selection for in-depth research, provide direction for more focused case studies and comparisons, and be used to provide additional tests of hypotheses generated by small *N* research. On the other hand, small *N* analyses can be used to assess the plausibility of observed statistical relationships between variables to develop better measurement strategies, particularly if the case studies are carried out in the "systematic" comparative mode suggested by George (1979). This integrated approach improves the prospects of making valid causal inferences in cross-national comparative research by drawing on the distinct strengths of the two approaches.

Utilizing the mixed approach in this study, we seek to maximize generalizability of findings using large *N* of UN missions in civil wars. We use the integrated method combining quantitative analysis and key representative cases of robust peacekeeping in Africa that include Sierra Leone and Democratic Republic of Congo, compared with some cases of traditional peacekeeping in Angola and Rwanda, to determine what might have accounted for the likely differences discovered in the quantitative analysis. In this way we maximize comparability by focusing particularly on one region, Africa, and contrasting two cases of robust with two cases of traditional peacekeeping. George's (2005) case study methodology will be employed, to answer the same set of research questions for each case. At the same time we will undertake a statistical analysis of all UN peacekeeping missions in civil conflicts between 1956 and 2006 in order to contrast the outcomes, in terms of effect on casualties and human rights of those with or without robust missions. Contextual variables will be those listed in the hypotheses of Chapter 2.

Unit of Analysis

Many scholars of peacekeeping use the mission as the unit of analysis. The disadvantage of using mission is that it may not reflect the true annual progression of events during the period of deployment. The reality is that changes in mission variables, such as troop size, may affect the number of civilians killed. Using the mission as a unit of analysis that compares the number of civilians killed at the start of the mission with that of the end of mission may miss the fluctuations between the starting date and the last date of the mission. In this study, the unit of analysis to be studied is the *mission year*. During civil wars civilian killings can occur in one year and not in another. Using mission *year* as the unit of analysis gives a true reflection of the number of civilians killed within the year, making it possible to account for possible reasons for fluctuations and progressions in civilian killings in different years. The yearly observation allows evaluation of the short-term effect of mission on civilian killings. It also permits us to systematically account for the variations in killings during the period of the mission. We could also determine whether robustness effects are dynamic; in other words, working for a while and then failing. Comparing civilian killings at the start of mission with killings at the end of mission does not reflect or account for civilian killings between the start of mission with killings at the end.

Data Collection and Variable Measurement

In order to test the hypotheses presented in Chapter 2, I generated a new dataset for all UN peacekeeping missions in civil conflicts from 1956 to 2006. The data was created using multiple sources. In this study, there are 46 individual missions (see table 2.1 Appendix 1) leading to 240 observations (mission-years) in the dataset covering 29 intrastate conflicts in 29 countries. In order to identify an event, a mission has to be deployed in civil war during the period of observation in accordance with the Uppsala Conflict Data Program (UCDP) definition of armed conflict; an incompatibility (over either governmental power or territory, or both) between a government and one or more irregular groups that in one year results in at least 25 battle-related deaths.[7] This threshold for inclusion is lower than in many studies of civil war that requires either 200 or 1,000 battle deaths a year[8] and also allows for a study of violence against civilians in low-intensity conflicts.

Dependent variable: civilian killings

The dependent variable is *civilian killing*. Following Valentino et al. (2001, 8), civilian is defined as "any unarmed individual who is not a member of a professional or guerrilla military group and who does not actively participate in hostilities by intending to cause harm to enemy personnel or property."

Civilian Killing is a count of the number of civilians deliberately killed by either the government of a state or by a formally organized nonstate groups, which results in at least 25 deaths in a year based on the UCDP *one-sided violence* dataset, which is an actor–year dataset with information on intentional attacks on civilians by governments and formally organized armed groups.[9] The program defines *one-sided violence as* the use of armed force by the government of a state or by a formally organized group against civilians that results in at least 25 deaths.[10] The UCDP, however, does not cover all my cases, so I supplemented my cases with data from United States Department of State annual country reports on human rights practices[11] and Human Rights Watch annual reports by country.[12] Only deliberate killings through the use of force are coded, which means that civilians killed in crossfire are excluded. Furthermore, indirect killings such as starvation of people in conflict areas are also excluded. These three sources of data were chosen for the dependent variable because they distinctively report on the actual number of civilians killed including both combatants and non-combatant civilians, unlike other data sources that only report "battle related" fatalities.[13] The state failure task force's project data on genocides and *politicides* also provides another source of civilian killings. The problem with this data source, however, is that the numbers of death are given as an annual number, coded as a scale from 0 to 256,000 deaths or more per year. For example, the number of civilians killed reported using a scale of ordinal numbers instead of actual number killed seems ambiguous. For example "0" represents less than 300 deaths in a year and "0.5" represents 300–1,000 deaths in a year. This scale seems ambiguous since it does not give a true reflection of the actual number of civilians killed in a year. The atrocities event data also reports on civilian killings but similarly uses the scale and also includes those injured.

Valentino et al. (2001) also provide data on mass killing of noncombatants but the limitation of these data is that it focuses only on state-sponsored killings excluding killings by organized nonstate actors that are parties to conflicts. It is only the UCDP that reports on civilian killings by both state and nonstate actors but it has certain limitations. The first limitation is that the data does not cover all my cases in this investigation, hence I supplemented my cases with data from the US State Department and Human Rights Watch annual country reports. Furthermore, there are several reliability problems of the data quality of civilian killings due to the limitation of the primary data sources. Civilian killings caused by states and non-organized groups are often underreported since neither actor wishes to publicize its atrocities and thereby jeopardize its reputation. Poor government record keeping, distant locations of atrocities, and poor communication also obstruct accurate report of civilians killed. For example, Rwanda in 1994 had conflicting reports of 500,000, 583,000, and 800,000 civilians killed (see UCDP dataset). We propose to get around some of these difficulties by including comparative statistics from more than one source.

Civkchang: This variable refers to changes in civilian killings—either decreased or increased civilian killings—to determine mission success or failure. This variable is coded as 0 = decreased civilian killings (mission success), 1 = increased civilian killings (mission failure).

Independent variables

In Chapter 2, I articulated eight core hypotheses to be tested in respect of the theses that robust peacekeeping works better than traditional peacekeeping in reducing civilian killings when missions are deployed in civil conflicts. The analysis includes the following independent variables: mission type, mission strength, arms type, major power participation, resources (funds), and rules of engagement. In what follows, I discuss measurement and coding of these variables (see Appendix 4 for the project codebook).

Mission type: The variable refers to the types of UN peacekeeping missions launched into civil conflict zones as detailed in Chapter 1 and these include observer, traditional, multidimensional, and "robust" peacekeeping missions. The type of mission is coded in the following manner:

1 = Observer mission
2 = Traditional mission
3 = Multi-dimensional mission
4 = "Robust" mission

Following Goulding (1994), I collapsed the missions and coded them as follows:

0 = Traditional mission (that includes missions 1, 2, and 3)
1 = "Robust" mission
Data source: Paul Diehl's peacekeeping data and UNDPKO

Mission strength/size: This variable is measured by the total manpower of the peacekeeping mission. Data for troops' manpower size is from UN Department of Peacekeeping Operations' website (http://www.un.org/Depts/dpko/ops.htm).

Rules of engagement: I conceptualize rules of engagement as directives issued by competent military authority (in this case the UN Security Council) that delineate the circumstances and limitations under which forces will initiate combat engagement with other forces encountered (US JCS publication 1). We measure rules of engagement by the provisions in the mandate on conditions for use of force. Data is from UN Department of Peacekeeping Operations' website. Rules of engagement are coded as follows:

0 = when force is not permissible to protect civilians from attacks but used only in self-defense
1 = when force is permissible to protect civilians from attacks

Arms type: It refers to the types of arms the peacekeeping contingents possess and use to fulfill the mandates. According to the United Nations, small arms, also known as light weapons, are conventional weapons that can be carried by a soldier or on a light vehicle. Small arms thus include revolvers and self-loading pistols, grenade, submachine guns, rifles, machine guns, mines and antitank weapons. The US army definition of small arms refers to guns/firearms less than 20 mm in caliber (Title 18 US Code 921). Heavy weapons are the major conventional arms as defined by the United Nations that include battle tanks, armored combat vehicles, artillery systems, combat aircraft, attack helicopters. Data is from the UN *Register of Conventional Arms*, SIPRI arms database. The variable arms type is coded as: 0 = small arms; 1 = heavy arms.

Resources: The total mission expenditure in US dollars. Data is from UN Department of Peacekeeping Operations' website http://www.un.org/Depts/dpko/ops.htm).

Major power: This variable refers to the major power contribution of troop contingents to the peacekeeping mission. In this dataset, the major powers indicate the five permanent members of the Security Council. Data for major powers' participation is from data for troops' manpower size from UN Department of Peacekeeping Operations' website (http://www.un.org/Depts/dpko/ops.htm). The major power participation is coded as follows: 0 = no participation of major powers in a peacekeeping operation; 1 = participation of at least one major power in a peacekeeping mission

Control variables

Ethnic affinity: This variable is a proxy for the cultural similarity between the mission and host country. Affinity between the field executive (e.g., the force commander) is measured using Vanhanen (1999) categorization of Caucasoid, Negroid, and Mongoloid. Source of data is UN Department of Peacekeeping Operations' website (http://www.un.org/Depts/dpko/ops.htm). Data on ethnic composition of countries is from *CIA World Factbook 2000* and Encyclopedia Britannica, while Cavalli-Sforza (1991) determines to which of the three main categories—Cascoid, Negroid, and Mongoloid—the ethnic groups belonged. Host country and country of origin of field executive are identified from mission reports of the Secretary-General, UN Department of Peacekeeping Operations' website. *Ethnic affinity* is coded as:0 = non-affinity; I = there is affinity.

Conflict type: This variable distinguishes identity conflict from ideological and revolutionary conflicts. The distinction is based on the coding rule of Regan (2002). The type of conflict is coded as 0 = ideological, revolutionary conflict; 1 = identity (ethnic or religious) conflict.

Border: This is operationalized in terms of either a shared border or less than 150 miles of water separating two contiguous land masses based on the measures used by the COW dataset (see www.crrelatesof war.org) continuity data. The number of borders is measured as a continuous variable that ranges from a low of one border to a high of nine borders.

Troop's composition: This variable is measured by the total number of countries contributing troop contingents to the peacekeeping mission. Data for troops' composition is from UN Department of Peacekeeping Operations' website.

Peace agreement: This variable is denoted by a dummy variable indicating whether peace agreement is signed or not by all principal combatants before peacekeepers are deployed. The data for the presence of peace agreement is from the peacebuilding dataset created by Doyle et al. (2000) or UN documents pertaining to the conflicts. The variable is coded as 0 = absence of peace agreement before deployment; 1 = presence of peace agreement before deployment.

Regime type: This refers to regime type, either democracy or autocracy, during the conflict whiles the mission is deployed. The source of data for this variable is the combined policy score collected in the Polity IV data. The democracy and autocracy scores are each ordinally scaled, ranging from zero to ten, each measuring institutional aspects of the regime. The regime type score therefore, ranges from −10 to 10.

Following Valentino et al. (2000), this variable is coded as a dummy variable: 0 = if the combined Polity score is less than +6 (less democratic); 1 = if the combined Polity score is equal to or greater than +6 (highly democratic).

Methodological Consideration

In order to explore the questions raised in Chapter 2 and to test the stated hypotheses, statistical analyses, using two different statistical models, will be utilized supplemented with qualitative analysis to examine the emergent conviction that "robust peacekeeping" missions work better than traditional peacekeeping missions in reducing human rights violations, especially civilian killings in conflict zones. I also examine key representative cases of robust peacekeeping in Africa that include Democratic Republic of Congo and Sierra Leone compared with some cases of traditional peacekeeping in Rwanda and Angola.

The most common model applied to event count in conflict research has been the ordinary least squares (OLS) regression analysis. In a recent study, Valentino et al. (2007) used OLS analysis for civilian killing, while a study by Erik et al. (2007) employed negative binominal regression analysis for one-sided civilian killing. Both studies show some disagreement about the choice of methodological models for event count (see King, 1989). According to King (1989, 845), OLS provides an unbiased linear estimator, and this is unaffected by different distributional assumptions but identifies certain problems in using OLS. First OLS assumes a linear relationship. This is an implausible functional because often it results in predicted events counts that are less than zero and therefore meaningless. Second, OLS takes into account neither heteroskedasticity nor the underlying Poisson distribution of the disturbances, hence OLS does not use all available information in the estimation. King (1989) and Long (1997) suggest that linear regression models may result in outcomes with inconsistent, biased, and inefficient estimates. These statistical problems of OLS could result in substantively biased conclusions rendering OLS a statistically inefficient model for event count.

The most common event count model, Poisson regression, determines the probability of an event count based on Poisson distribution. Unfortunately, Poisson regression models may bias standard errors downward, leading to inaccurate assumptions about the level of significance of independent variables. In addition, Poisson regression also assumes that the mean distribution is equal to the variance making the resulting parameter estimates inefficient when a researcher suspects overdispersion.[14] Overdispersion or variance greater than the mean is quite likely in political science research.[15]

Given that the dependent variable is a count of civilians killed, the hypotheses are tested applying negative binomial regression model (NBRM) as argued by Gary King (1999) and used by Erik et al. (2007). An alternative approach to NBRM will be the Gamma count model.[16]

An NBRM is suitable because it is preferred when there is overdispersion in the data, which indicates that there might be both contagion and unobserved heterogeneity.[17] Apart from unobserved heterogeneity and contagion, which are defined through the overdispersion in the data, there are also many zeros, and NBRM predicts a higher number of zeros.[18] This seems reasonable since there may be fluctuations in civilian killings during the mission period. Within a mission year there may be civilian killings while in other years they may not. To determine either mission success or failure, I employ logit regression model given that the dependent variable is dichotomous. The analysis will allow us to understand how the number of civilian killings in a mission year is affected by the independent variables controlling for the intervening variables. Results of the quantitative analyses are reported in Chapter 4.

Borrowing from George's (1979) structured, focused comparison model, I examine in Chapter 5 key representative cases of robust peacekeeping in Africa that include UN missions in Sierra Leone and the Democratic Republic of Congo as well as traditional peacekeeping missions in Angola and Rwanda in the context of both the dependent variables and control variables to determine what might have accounted for the likely differences discovered in Chapter 4. I also explore other potential factors not included in the statistical model that might affect mission outcome. These representative cases were chosen from the same region, in order to minimize cultural and other biases, and they were chosen because the excesses in terms of human rights violations in these conflicts attracted both scholarly and policy makers' attention. They also share similarities with slight differentiations that will be discussed in detail in Chapter 5.

Conclusion

The methodology employed in the study not only helps us to reach a better understanding of how civilian killings during UN missions is associated with mission type, given the lack of research on robust mission, but also contributes to the theoretical understanding of the factors contributing most to civilian killings during UN missions. The study should allow policy and theoretical conclusions by policy makers about these factors in an attempt to realize the cherished objective of the protection of human rights in civil conflict zones. Policy implications will be discussed in Chapter 7.

5

Statistical Analysis of Robust Peacekeeping and Civilian Protection in Civil Wars

This chapter details the results of the statistical analyses of the newly constructed dataset. As discussed in Chapter 2, this study argues that robust peacekeeping missions by the United Nations are more likely effective than traditional peacekeeping missions by United Nations in reducing civilian killing in civil conflicts. The methodology described in Chapter 3 provided the means to analyze the relationship statistically.

Findings

Descriptive statistics

For the period 1956–2006 that this study covers, the United Nations deployed 46 missions in 29 civil conflicts. In terms of mission years, the United Nations engaged in 240 mission years during the study period. The data depicts that within this period, the percentage mission years for the mission types are: observer, 33 percent; traditional, 28.3 percent; multidimensional, 18.3 percent; robust, 20 percent (see table 5.1 for percentage mission years). In our estimation, the lowest number of civilians intentionally killed by conflict parties in a given year was 25 with 500,000 as the highest recorded in the Rwandan genocide. The data depicts that 97.7 percent of civilian killings are during traditional peacekeeping missions including observer missions, traditional, and multidimensional peacekeeping, while robust peacekeeping registers only 1.3 percent of civilian killings (see table 5.2).

Table 5.1 Percentage years of UN mission types, 1956–2006

Mission type	Number of mission years (percentage)
Observer mission peacekeeping	80 (33.3)
Traditional peacekeeping	68 (28.3)
Multidimensional peacekeeping	44 (18.3)
Robust peacekeeping	48 (20)
Total	240 (100)

Table 5.2 Percentage intentional civilian killings of UN mission types, 1956–2006

Mission type	Civilian killings (percentage)
Observer mission peacekeeping	3,916 (.7)
Traditional peacekeeping	501,923 (95)
Multidimensional peacekeeping	11,161 (2)
Robust peacekeeping	7,073 (1.3)
Total	524,073 (100)

Statistical Results

To answer the research questions and test the hypotheses in Chapter 2, both negative binomial regression and Logit regression models have been used (see table 5.3 Appendix E for the correlation matrix). Table 5.4 presents the results of the negative binomial models of intentional civilian killings in UN missions in civil wars between 1956 and 2006. Model 1 tests our basic hypothesis excluding the control variables. In order to test the robustness of the model, Model 2 contains all relevant control variables discussed in Chapter 3.

Results of model 1 in table 5.4 support our core hypothesis that robust peacekeeping is more likely to be associated with lower number of civilian killings. The variable robust peacekeeping has a negative and statistically significant association with the number of civilian killings, meaning robust peacekeeping missions are more successful in lowering the number of deliberate civilian killings by conflict parties in civil conflicts. The mission strength variable also has a negative and statistically significant relation to

Table 5.3 Negative binomial regression models of intentional civilian killings in United Nations Missions in civil wars, 1956–2006

	Model 1 Coefficients of the number of intentional civilian killings (standard errors)	Model 2 Coefficients of the number of intentional civilian killings (standard errors)
Robust peacekeeping	−2. 287** (0.922)	1.166 (0.831)
Mission strength (1,000 people)	−0. 140*** (0.030)	−. 200*** (0.032)
Resources (in US$1,000)	4.860 (6.140)	3.850 (3.110)
Major power participation	4.223*** (1.054)	0.209 (0.726)
Rules of engagement	1.020** (0.300)	−1.106 (2.528)
Ethnic affinity		−0.081 (0.702)
Identity conflict		4.280** (1.352)
Borders		0.407 (0.393)
Democracy		0.151 (0.750)
Troops composition		0.057** (0.023)
Constant	4.100*** (0.256)	1.415 (1.558)
Number of observation	197	144
Log pseudo-likelihood	−870.573	−724.046
Wald χ^2	30.01	96.92
Prob > χ^2	0.000	0.000

Note: Estimations have been performed using Stata 9. Robust Standard errors are in parentheses (adjusted for clustering on missions). $^{*}p < 0.1$, $^{**}p < 0.05$, $^{***}p < 0.01$. All tests are two-tailed.

the number of civilian killings, confirming hypothesis 1a. This implies that a large force strength is associated with lower levels of the number of civilian killings. This could mean that such forces shorten the wars, or protect the populations more effectively, or intervene in wars that have low casualty rates to begin with.

Interestingly enough and contrary to expectation, however, major power participation has a positive and statistically significant relationship to the number of civilian killings at least under model 1, implying that major power participation in UN peacekeeping operations is more likely to be associated with higher numbers of civilian deaths tending to disconfirm hypothesis 1e. Similarly, the rules of engagement variable has a positive and statistically significant effect in model 1, indicating that deliberate civilian killings likely to be higher when the peacekeeping mandate permits mission forces to use force other than for self-defense, disconfirming hypothesis 1d. These findings might give pause to easy assumptions that involving major powers is a reliable way to stabilize local violence or

Table 5.4 Marginal effects of statistically significant variables on intentional civilian killings in UN Missions in civil conflicts, 1956–2006

	Model 1 Approximate change in the number of intentional civilian killings	Model 2 Approximate change in the number of intentional civilian killings
Robust peacekeeping	−1,442	
Mission strength (1,000s)	−142	−32.2
Major power participation	1,611	
Rules of engagement that permits the use of force other than self- defense only	1,681	
Identity conflict (ethnic and religious)		4,240
Troops composition comprising many culturally diverse contingents		10

Note: Marginal effects have been computed with the coefficient from the negative binomial regression models presented in table 2. The values of selected statistically significant variables changed while holding all others constant at their means or modes.

that broad latitude in use of force by peacekeepers is reliable for purposes of civilian protection. This is explained by the fact that spoilers may strategically label the involvement of major powers imperialistic, which would help mobilize support and heighten their resistance against the imperialists. When the United States clamored for a robust mission in Darfur, the Sudanese president Al Bashir criticized the deployment of UN force as a foreign invasion, calling on Sudanese and Muslims to "rally around the flag" to defend Sudan's national sovereignty. President Al Bashir also tactically labeled the deployment of a US-supported robust UN mission imperialistic and yet another US attack against an Islamic state as it did in Iraq and Afghanistan. He vowed to resist ferociously against the imperialists and called on all Muslims to join forces and resist US attack on Islamic values. Obviously, had the United Nations deployed a robust mission in Darfur without addressing the concerns of the Sudanese government, the UNAMID would have been resisted and both mission's forces and civilians would be attacked resulting in increasing levels civilian killings. The lengthy discussion between the United Nations and the Sudanese government to avert an imminent confrontation and forceful resistance delayed the deployment of UNAMID and also handed the mission a watered-down robust mandate that was incapable of protecting civilians. The United Nations' response to Darfur killings will be discussed in detail in Chapter 6. The variable resource (funds) has no significant effect.[1] These variables will be further explored in the next chapter utilizing a case study analysis.

Turning to the results in model 2 that includes the control variables, we first notice that mission type itself is no longer negatively and significantly related to casualty levels, while mission strength as in model 1 has a negative and statistically significant relationship with civilian killings, further confirming hypothesis 1a. Robust rules of engagement have a negative effect on civilian killings, thus it lowers civilian killings but it is statistically insignificant. The findings also support the argument in the deterrent model that large troop deployments increase the costs of continued fighting for combatants and thereby reduce civilian killings. This is possible since a large force can effectively monitor a large area and pose a formidable deterrent to spoilers. However, none of our other core variables had significant effects on the number of civilian killings. Major power participation has positive links to civilian killings but not the statistically significant ones.

Examining the control variables, identity conflict has a positive and significant effect on civilian killings tending to confirm hypothesis 5. Thus, identity conflicts are more likely to be associated with higher numbers of civilians killed during internal conflicts than are ideological forms of dispute. The latter point supports the argument that identity conflicts are so much based on deep-rooted emotional values that combatants find it difficult to compromise on those underlying issues and restrain forces. In view of this realization, violent wars become the most viable mechanism of the pursuit of group objectives. It therefore presupposes that casualties will be more difficult to control in identity conflicts than in ideological conflicts similar to the Rwandan genocide. Peacekeeping contingent composition also is positively and significantly related to civilian killings confirming hypothesis 3. This finding is in line with the argument that the more the mission force comprises many culturally diverse contingents, the less likely the mission success because cultural differences among peacekeeping forces may result in disagreement in interpretation of the resolution and what actions to take (Elron et al., 1999; Duffey, 2000). Such disagreements on the interpretation of the resolution and rules of engagement may delay or prevent actions to protect civilians. For example, during the early stages of UNAMSIL's operation in Sierra Leone, the disagreement between the Indian command and Nigerian contingents paralyzed the mission and adversely affected to some extent the mission's operational efficiency (Bullion, 2001, 78). All the other control variables have no significant effect on civilian killings.[2]

Table 5.5 Logit regression model of intentional civilian killings in United Nations Missions in Civil Wars, 1956–2006

	Coefficients of change in the number of intentional civilian killings (standard errors)
Robust peacekeeping	−2.088*** (0.594)
Major power participation	−1.157 (0.594)
Mission strength (1,000)	0.000 (0.000)
Resources (US$1,000)	7.850 (7.790)
Identity conflict	2.164** (0.844)
Borders	0.135 (0.388)
Troops composition	0.054** (0.027)
Ethnic affinity	1.520* (0.781)
Democracy	1.435** (0.727)
constant	1.99217 (1.4501)
Number of observation	102
Log pseudo-likelihood	−45.702813
Pseudo-R^2	0.2704

Note: Estimations have been performed using Stata 9. Robust Standard errors are in parentheses (adjusted for clustering on missions). $^{*}p < 0.1$, $^{**}p < 0.05$, $^{***}p < 0.01$ All tests are two-tailed.

Table 5.5 presents marginal effects of statistically significant variables of the negative binomial model on civilian killings in UN peacekeeping missions in civil conflicts from 1956 to 2006.[3] These marginal effects offer us more substantive insight into the extent to which robust peacekeeping, mission strength, the use of force, major power participation, identity conflict, and troops composition influence civilian killings during the period under investigation. According to model 1, with robust peacekeeping, about 1,442 fewer civilians are killed per year. In respect of mission strength, we find that an increase in the mission strength by one standard deviation from the mean lowers the number of intentional civilian killings by 142 per year. Considering other factors, we find that conflict parties kill 1,611 more civilians per year with major power participation in UN missions in civil conflicts. Furthermore, we find that 1,681 more civilians' deaths per year occur when the mandate permits UN mission forces to use force for purposes other than self-defense only.

In respect of model 2, which included our control variables, we do not have these latter cautionary effects, and we find that an increase in the mission strength by one standard deviation from the mean lowers the number of intentional civilian killings by 32 per year. Considering other factors, we find that in identity conflicts, 4,240 more civilians' deaths per year can be expected than in other forms of conflict. In terms of troops' composition, we find that an increase in troop diversity by one standard deviation from the mean, results in roughly 10 more civilian deaths per year.

Table 5.5 presents the results of a logit regression model for our binary dependent variable change in civilian killings coded as "0" for mission success or reduced deliberate civilian killings and "1" for mission failure or increased deliberate civilian killings.[4] Robust peacekeeping has a negative and significant relation to civilian killings, suggesting that robust peacekeeping lowers civilian killings in line with the findings in model 1 table 4.4 and again confirming the core hypothesis 1. The rest of the independent variables, however, appear to have no significant impact on civilian killings. Turning to the control variables, conflict type, troops' composition, ethnic affinity with the host state and democracy have positive and significant connections to civilian killings. In other words, identity conflicts and culturally diverse troop contingents and even somewhat democratic states are more associated with civilian killings confirming our earlier findings in model 2 table 4.4. Ethnic affinity of the mission force presents a very interesting finding. The results show that the forces' affinity with the conflict state may increase civilian killings defying the basic argument for regionalism as a better mechanism for peace operations. Since the mission force share the same culture with the conflict parties, they may have interest in the conflict thereby compromising the principle of impartiality and hurting a party to the conflict, ultimately obstructing the peace process.

Also, the cultural bond between the mission force and conflict parties, along with concerns about being accused of partiality, may hinder the mission force undertaking certain stringent measures against spoilers of the peace process. However, the data leaves us merely to surmise such possibilities, pending our case study analysis. Finally, contrary to expectation, the finding suggests that democracies are associated with more civilian killings. This finding, however, might confirm the Davenport et al. (2004) discovery that there is a threshold of democratic peace and that lower levels of democracy or transitional democracies are associated with human rights violations (Davenport et al., 2004, 551). Nascent democracies can be among the most nascent violent forms of statehood.[5] This is the stage of hybrid of or anocratic regimes characterized as the center of the democratic continuum. Anocracies are situated midway between an autocracy and a democracy. At this stage of democratic metamorphosis, the state emerging from authoritarianism and in the process of democratization has not built developed democratic institutions. In this period we still witness remnants

Table 5.6 Marginal effects of statistically significant variables on intentional civilian killings in UN Missions in civil conflicts, 1956–2006

	Approximate percentage change of probabilities of intentional civilian killings
Robust peacekeeping	−0.31 (31%)
Democracy	0.29 (29%)
Troops composition	0.01 (10%)
Identity conflict	0.40 (40%)
Ethnic affinity	0.24 (24%)

of the old regime violently opposed to change to democracy and power is spread amongst elite groups who are constantly competing with each other for power. Anocracy is the least resilient political system to short-term shocks. It promises an inclusive and effective socio-politico-economic system of governance. However, by its fundamental philosophy and designation, it threatens members of the established elite; and is therefore very vulnerable to disruption and armed violence by the supporters of the status quo. Examples of anocracies in Africa include the warlords of Somalia and the shared governments in Kenya and Zimbabwe.

Table 5.6 presents marginal effects of statistically significant variables on civilian killings of the logit model. Here we see that, robust peacekeeping decreases civilian killings by about 31 percent per year. In respect of the control variables, we find that democracies increase civilian killings by 29 percent per year and an increase in troops' diversity by one standard deviation from the mean increases civilian killings by 10 percent per year. Finally, we find that identity conflicts increase civilian killings by 40 percent per year and ethnic similarity to host state also increases civilian killings by 24 percent per year.

Conclusion

The empirical results presented in this study support the core argument that robust peacekeeping works better than traditional peacekeeping in reducing civilian killings. This finding provides support for the theory of force deterrence, and that a large robust UN force deters spoilers from killing civilians. A unique finding is the contribution of large strength to mission success. This resonates with the Brahimi Report's recommendation for large troop size for mission success because large troops may occupy large area effectively monitoring and implementing mission mandate and posing barriers of one sort or another to war-mongers. (United Nations, Report of the Panel on United Nations Peace Operations, 2000). However, the related notions of major power involvement and more militarily assertive missions or rules of engagement do not necessarily appear to restrain killing and in fact are associated in some analyses with significantly higher civilian casualties. This is explained by the fact that major power involvement is tactically framed as imperialistic to rally the locals round the flag and defend their values against the aggressor or neocolonialist. The findings further inform us of the severity of identity conflicts and confirm the argument that highly culturally diverse peacekeeping forces can derail mandate implementation as in the UNAMSIL experience in Sierra Leone (Bullion, 2001). The finding that regional cultural similarity results in mission failures is striking since it defies the popular assumption by UN policy makers that regionalism is the better option for international security due to forces' similarity with neighbors. We also observe that borders matter and countries bordered by many states may be prone to increased civilian killings in civil war situations. It was also found that at least proto-democratic status of the host state appears to be the harbinger of a potentially higher total deliberate civilian killing. The findings thus have both theoretical and policy implications in the field of peacekeeping. Perhaps robust peacekeeping per se may not be enough to stave off deliberate civilian killings. Robust peacekeeping may be affected by late deployment, external interventions, vastness of the host country, the number of borders, arms shipment and the number, the willingness to fight and the cost-tolerance level of the spoilers as well as the mission force as exemplified by the robust mission (MONUC) in the Democratic Republic of Congo. In view of the findings and the need for more refined understanding of the process at work in pacification through peacekeeping, we examine in Chapter 6 key representative cases of robust peacekeeping in Africa that include Democratic Republic of Congo and Sierra Leone compared with some cases of traditional peacekeeping in Rwanda and Angola to determine what might have accounted for the differences and ambiguities discovered in Chapter 5.

6

Case Studies: Testing the *Formidable Barrier* Model of Robust Peacekeeping Success

The quantitative analyses presented in Chapter 5 indicate results contrary to our expectation in terms of the impact of certain salient constituent elements of robust peacekeeping such as robust rules of engagement and major power participation on lowering civilian killings. Robust peacekeeping per se may not be enough to stave off deliberate civilian killings and could be limited by certain operational and structural constraints. The effectiveness of Robust peacekeeping could be hampered by late deployment, major power neorealist interests in conflict, external parties or states interventions or involvement in conflict, vastness of the host country, the number of borders, arms shipment and the number, the willingness to fight and the cost-tolerance levels of the spoilers as well as the mission force. A unique finding is that a large military force reduces intentional civilian killings in civil conflicts while the other core factors theoretically associated with effectiveness of robust peacekeeping have intangible effect. I also find that identity conflicts, lower levels of democracy in the target state, a more culturally diverse mission force, and a mission force similar in culture with the host state correspond to increased intentional civilian killings. Yet due to various problems such as coding and measurement decisions and the temporal order of variables, statistical analysis alone may not be able to capture the true impacts of such factors. A more detailed study of individual cases would help establish the actual factors that impacted civilian killings. A closer look at individual cases therefore serves to inform the author as well as the reader of other potential significant factors impacting on civilian killings.

In this chapter, I examine key representative cases of robust peacekeeping in Africa that include UN missions in Sierra Leone and the Democratic Republic of Congo as well as traditional peacekeeping missions in Angola and Rwanda to determine what might have accounted for the differences discovered in Chapter 5. These representative cases were chosen from the same region, in order to minimize cultural and other biases, and they were chosen because the excesses in terms of human rights violations in these conflicts attracted both global scholarly and policy makers' attention. They also share similarities with slight differentiations. The Angolan and Rwandan cases, both instances of traditional peacekeeping, were both characterized by limited strength of troops, the use of mainly small arms, and limited rules of engagement. A striking difference between these two cases is that while major powers had interest in the Angolan case, they did not have similar political involvements in the Rwandan case. In the Angolan peacekeeping case for the period 1988–1999, the United States and the Soviet Union had stakes in the conflict and therefore covertly influenced both the war (especially through arms supplies and support for favored factions) and the peacekeeping mission in the sense that their support for the conflict parties prolonged the conflict with resultant civilian killings. Their apparent disinterest in the Rwandan mission led to a poorly designed traditional peacekeeping mission reflective of its small troop size of about 2,500, poor and small weapons, poor resources, and limited rules of engagement that did not mandate the United Nations Assistance Mission for Rwanda (UNAMIR) forces to protect civilians under threat, contrary to what was agreed on by the conflict parties in the Arusha Accords, which led to a genocide.

The UN missions in both Sierra Leone and Democratic Republic of Congo (DRC) were cases of robust peacekeeping that had traditional missions from their inception and later transformed into robust missions due to failures to protect UN personnel and civilians. However, while the Sierra Leonean mission (UNAMSIL) succeeded in protecting civilians, DRC's MONUC has produced mixed results of civilian protection.

I examine these cases testing the *formidable barrier* model with two goals in mind. The first is to tease out the causal relationships between key independent variables that include mission type, mission strength, rules of engagement, weapon types, major power participation, and intentional civilian killings. Additionally, I examine whether the cost/benefit calculations affected conflict parties' decisions to either continue to fight or choose peaceful settlement. The second goal is also to ensure that the control variables and other potential factors equally critical to deliberate civilian killings are not overlooked. The cases are examined below using a structured case analysis methodology.[1]

United Nations Angola Verification Missions (UNAVEM) I, II, III/United Nations Observer Mission In Angola (MONUA)

The Angolan civil war

Angola was engulfed in war since the mid-1960s: first its war of independence, and then a bloody war fought from 1975 to 2002. The 27-year-old civil war was the product of Angola's violent internal dynamics and massive foreign intervention. The civil war was primarily a struggle for power among the three former liberation movements, the Popular Movement for the Liberation of Angola (MPLA), led by Agostinho Neto; the Angola National Liberation Front of Angola (FNLA), led by Holden Roberto; and the National Union for the Total Independence of Angola (UNITA), led by Jonas Savimbi due to personal jealously, contrasting ideologies, and ethnic animosities. The ethnic divisions that divided the parties were compounded by the Cold War ideological and geostrategic cleavages of the West and the East. The conflict duration, intensity, and severity were heightened with the involvement of both the West and the East that considered the conflict to be critical to the global balance of power and to the outcome of the Cold war. The major power blocs made it a proxy war between them informed by Cold War neorealist geostrategic considerations, which ultimately became one of the bloodiest wars of the Cold War era.

Three nationalist groups emerged to organize the fight for independence from Portuguese rule in the 1950s and 1960s. However, these anticolonial movements were divided along tribal, ethnic, and ideological lines with different regional origins. These differences prolonged the fight into the post-independence era. On January 15, 1975, following Angola's independence, the three parties signed the Alvor Agreements, in which they agreed to set up a tripartite coalition government for the fledging nation. However, personality, ideological, and ethnic conflicts arose. Shortly after the installation of the new government on January 13, 1975, the agreement broke down as fighting erupted among the parties. The MPLA, with the support of a significant 12,000 Cuban troops and Soviet guidance and military equipment worth US$200 million, defeated the forces of FNLA and UNITA and formed a government in 1976 that was recognized by the United Nations as the legitimate representative of Angola. What followed was a devastative Cold War proxy civil war involving the two major power blocs.[2] The MPLA received support from the USSR and Cuba; the FNLA was directly assisted by the US intelligence services, received arms and advisers from China, and received military aid from Zaire; and UNITA received backing from South Africa and later from the United States and from Portuguese settlers. With the extinction of

FNLA, the conflict continued between the United States-/South Africa-backed UNITA and the Soviet-/Cuba-backed MPLA. The Angolan and American governments began negotiating in June 1987.[3] All conflict parties joined the negotiations in New York and Geneva and agreed to a ceasefire on August 8. Representatives from the governments of Angola, Cuba, and South Africa signed the New York Accords in New York City on December 22, 1988, granting independence to Namibia and ending the direct involvement of foreign troops in the civil war.[4] Following the New York peace agreement, the United Nations Security Council passed resolution 626 creating the United Nations Angola Verification (UNAVEM), a peacekeeping initiative that started deployment in Angola in January 1989.[5] There was a ceasefire but it broke down and hostilities resumed.

As a result of a complex international diplomatic process and the essential defeat of South African and insurgent forces at the hands of the Cuban-backed MPLA government, the conflict parties reached agreement on both the implementation of Security Council resolution 435 (1978), leading to the independence of Namibia, and Cuban troops began withdrawal from Angola in 1989 and by 1991 they were gone completely.[6] Under pressure from the United States and the Soviet Union, President José Eduardo dos Santos of Angola and Savimbi signed the Bicesse Accords in Lisbon, Portugal, on May 31, 1991, with the mediation of the Portuguese government. The accord called for an immediate ceasefire, demobilization, and return of both armies to cantonment, creation of a national army, and free and fair elections in September 1992. United Nations UNAVEM II mission was mandated to supervise the elections but UNITA largely did not disarm. UNITA complained that its supporters were not being registered. Despite protestations from the United States, NGOs, and other parties, the MPLA unilaterally ended the registration process on August 1992. The first round of the presidential elections was held on September 29–30 but none of the candidates received 50 percent or more of the votes, thus requiring a second round of voting between the top two contenders in accordance with the election law. UNITA, along with other opposition parties and many other election observers, complained about a lack of transparency and fraud.[7] The election process broke down on October 31, when government troops in Luanda conducted a pogrom against UNITA in which civilians were killed and hundreds of UNITA supporters were detained. In a few days, the MPLA massacred over 10,000 UNITA and FNLA voters nationwide in what became known as the "Halloween Massacre."[8] Following this, Savimbi refused to negotiate for the electoral runoff and resumed the armed struggle against the MPLA. By August 1993, UNITA had gained control over 70 percent of Angola, but the government's military successes in 1994 forced UNITA to seek peace and the Lusaka Protocol was signed in Lusaka, Zambia, on November 1994, which reaffirmed the Bicesse Accords.[9] However, mutual distrust between UNITA and the MPLA, loose international oversight, the importation of foreign arms, and an overemphasis on maintaining the balance of power led to the collapse of the protocol.[10] While fulfilling most of the Lusaka Protocol, UNITA was hesitant and slow to return areas under its control to the government. In December 1999, President dos Santos declared that the only way to attain definitive peace was to isolate Savimbi and UNITA domestically and internationally, as well as neutralize him politically and militarily. He believed that the only way to achieve peace was through war and asked MONUA to leave. UNITA took it as a declaration of war, opening the way for return to hostilities.[11] On February 22, 2002, Jonas Savimbi was killed in a battle with government troops, compelling the UNITA rebels to seek peace, which led to the signing of the *Luena Memorandum of Understanding* as an addendum to the Lusaka Protocol in Luena on April 4. Since April 2002, Angola has experienced relative peace but it is a negative peace because the civil war caused more than half-a-million deaths, displaced 3.5 million poverty-stricken persons, and left a devastating legacy of landmines that continue to kill and maim innocent civilians to this day.[12] The war devastated Angola's infrastructure, and dealt a severe damage to the nation's public administration, economic enterprises, and religious institutions.

UNAVEM I: 1988–1991

The establishment of UNAVEM I in 1988 was the sequel of a complex international diplomatic process and the defeat of South Africa by Cuban-backed MPLA that culminated in Geneva and New York agreements between the conflict parties on both the implementation of Security Council resolution 435 (1978) leading to the independence of Namibia, and the withdrawal of Cuban troops from Angola in the 1990. As part of the New York Accords in New York City on December 22, 1988, both MPLA and UNITA requested the UN Security Council to verify the withdrawal of the 50,000 Cuban troops. Subsequently, the first United Nations Angola Verification Mission (UNAVEM I) was established by UN Security Council resolution (SCR) 626 (1988) of December 20, 1988, and deployment began in 1989 to oversee and verify the withdrawal of the 50,000 Cuban troops from Angola. With a force strength of unarmed 70 military observers without major power participation and no mandate to protect civilians under imminent threat, the mission successfully achieved its mandate but could not stave off further deliberate killing of civilians and several more years of sporadic inter-factional fighting in Angola since it lacked the capacity to do that.[13]

UNAVEM II: 1991–1995

Following the withdrawal of the Cuban forces and the Angola Peace Accord (Bicesse Accord) signed by the conflicting parties on May 31, 1991, UNAVEM II was established by Security Council resolution 696 (1991). Composed of 350 unarmed military observers, the mission was deployed with a mandate to verify the ceasefire arrangements throughout the country, to monitor troop assembly areas and neutrality of the Angolan police, and to observe and verify elections in accordance with the Peace Accords. The accords preceding the mission eased up tensions in 1991 as both parties agreed to hold elections in 1992 to be monitored by UNAVEM II. However, due to UNAVEM II's poor capacity due to its limited strength and resources, elections monitors were few and inefficient.[14] Apart from the problems of limited strength and other resources to effectively monitor the elections, UNAVEM II was also accused by UNITA of favoring MPLA given the United Nations' inimicalness toward UNITA. UN personnel reportedly were far from objective and refused to investigate UNITA claims of massive electoral abuses by the MPLA. Consequently, UNITA predictably did not cooperate with UNAVEM II, rearmed, and resumed the conflict, leading to the breakdown of a 16-month ceasefire in late October 1992. UNITA launched a nationwide operation to occupy municipalities by force and remove the government's local administrative structures raising tension in Angola resulting in increased deliberate civilian killings in 1992.[15]

The Security Council expressed concern at the rising tension and called upon both parties to fulfill their commitments under the Peace Accords, in particular the confinement of their troops and collection of weapons, demobilization, and the formation of the unified Angolan Armed Forces. It requested UNITA to respect the results of the elections and urged the leaders of the two parties to engage in immediate dialogue to enable the second round of the presidential elections to be held but UNITA continued its military action. The Security Council by resolution 864 (1993) condemned UNITA for its continuing military actions and, acting under Chapter VII of the Charter, imposed an embargo on the supply of arms and petroleum products to UNITA and restrictions on the travel of UNITA personnel. The Council insisted that UNITA respect the 1991 Peace Accords, and that the parties make every effort to restart negotiations.[16]

Following UNITA's resumption of the conflict, the MPLA reacted with summary execution of UNITA supporters even in the presence of UNAVEM II forces who could not stop the killings in view of its operational limitations. The conflict escalated in 1993, resulting in increased intentional killing of civilians as the accord collapsed and conflict engulfed many towns.[17] The small strength of UNAVEM II was confined to the capital and could neither deter nor stop the execution of civilians since it lacked the strength, appropriate rules of engagement, heavy weapons, and major power participation.

UNAVEM III: 1995–1997

The United Nations continued its efforts to facilitate the resumption of the peace process in consultation with the Angolan parties and interested countries, including, in particular, the observer States to the Peace Accords—Portugal, the Russian Federation, and the United States. On November 15, 1993, following extensive consultations and exploratory talks, the Government of Angola and UNITA began talks in Lusaka, Zambia. The question of national reconciliation proved to be the most difficult, since it involved matters such as the allocation of senior government posts to UNITA, including the governorships of provinces.

After laborious negotiations, an agreement was reached in October 1994 and the Lusaka Peace Accords was signed by both parties to restore peace. UNAVEM III was subsequently established by Security Council resolution SCR 976 of February 9, 1995, with an expanded lightly armed force strength of 4,222 (7,000 at its maximum strength) and a mandate to assist the parties to restore peace and achieve national reconciliation on the basis of the Peace Accords, the Lusaka Protocol, and relevant Security Council resolutions.[18] The peace accords coupled with increased force strength reduced fighting. Ceasefires were, however, frequently violated and both used this time to rearm. Despite a generally improving situation, conditions were volatile and hostilities were renewed in December 1995 with civilian killings but the conflict eased again early January 1996 reducing deliberate civilian killings in 1996.[19]

As a result of the mediatory role of third parties including South African president Nelson Mandela, the parties agreed to form a national government and a government of Unity and National Reconciliation was formed in 1997.[20] However, both parties were still far apart in meeting elements of the Lusaka Protocol concerning the demobilization of armed forces and formation of a joint military. There was some progress in the peace process in 1997 when Savimbi sent 63 UNITA deputies to take their seats in the National Assembly in Luanda.

MONUA: 1997–1999

Nevertheless, the West, which was unwilling to support the large UNAVEM III force indefinitely, seized on reports of progress to reduce UN presence in Angola to a smaller 1,158 and further reduced to a 396-man observer force, renamed the United Nations Observer Mission in Angola (MONUA) by Security Council resolution 1118 (1997) of June 30, 1997, to replace UNAVEM III. MONUA, reduced to a 396-man observer force, was mandated, in addition to the provisions of the UNAVEM III, to help the parties in consolidating peace and national reconciliation, enhancing confidence building, and creating an environment conducive to long-term stability, democratic development, and rehabilitation of the country.[21] The mission was to allow for the repatriation of a number of peacekeepers. Unfortunately, this was not to be. The remaining tasks were not accomplished and the country slid back into war as UNITA failed to comply fully with its obligations under the Lusaka Protocol with the justification that the government instead of demobilizing its troops as per the accord was rather transforming its troops into the secret riot police. The Security Council on June 12, 1998, passed resolution 1173 imposing Chapter VII sanctions against UNITA for resisting implementation of the Lusaka Protocol. The resolution froze UNITA's assets and barred foreign officials from meeting UNITA representatives in areas of Angola not under the control of the government. Realizing the international diplomatic isolation of UNITA and MONUA's reduced strength, MPLA launched massive attacks on UNITA and its supporters, which led to increased intentional civilian killings.[22] With the deterioration of the security situation in various parts of the country in 1999, conflict resumed and the security situation became more precarious. MONUA, due to its poor strength, limited mandate, small and insufficient arms, and lack of major power participation, could not restore peace, so the Security Council ended the mission in Angola on February 26, 1999, by Security Council resolution 1229 (1999).

Fuelled by the growing oil revenues, mercenaries, and political support from the Clinton administration, the MPLA gained a decisive advantage over UNITA by 2001. On February 23, 2002, MPLA troops killed UNITA Leader Jonas Savimbi ending the war in Angola in April 2002, followed by a ceasefire and demobilization of the UNITA forces.

Analyses

UN missions in Angola made a major contribution to the impressive achievement represented by the successful, internationally monitored conduct of elections in a war-torn country. After the renewal of hostilities, UNAVEM II maintained its political and military presence and was an essential factor in United Nations' effort to facilitate the resumption of negotiations to advance the peace process as well as in monitoring the evolving precarious situation in the country. UNAVEM II was an indispensable channel for communications and repeatedly drew the warring parties back to the negotiating process while fulfilling other vital functions, such as its support for humanitarian activities. To some extent, the mission became important as a preventive measure to check further escalation.

However, the setbacks experienced by the mission show difficulties the United Nations faces when its mandate and resources are inadequate in relation to the complexities of the task in a volatile conflict, especially in circumstances where the parties do not demonstrate the necessary political will for peace. UN peacekeeping missions in Angola failed to protect civilians from being killed by the conflict parties because its designation in traditional terms and limited resources undermined its capacity to constitute a formidable barrier to the violators. Also, the nondemocratic political system in Angola contributed to mission failure to protect civilians due to the absence of rule of law and democratic accountability. The *formidable barrier* model of peacekeeping argues that, the mission's capacity to protect civilians hinges on its ability to raise the cost of continued fighting comparative to the choice of peace. The Angola mission's incapacitation to raise cost of fighting for the belligerents are attributable to its limited resources, force size, nonuse of weapons or use of small arms, and limited rules of engagement often not including protection of civilians.

Other factors excluded from the statistical model in Chapter 5 but affected intentional civilian killings during the Angolan mission are also captured in the analyses. These factors are the supposed partiality of the UN mission, poor topography, late deployment, lack of political will of the conflict parties and covert major power interest involvement in the conflict, and continued arms supply.[23] The resources given to the missions were inadequate even for its limited monitoring role. Without adequate personnel and resources, it could not perform even the simple role of monitoring let alone the difficult task of deterring deliberate civilian killings. The problem of resources was summed up by the mission's Special Representative to the Secretary-General (SRSG) Margaret Anstee's observation that "we have been asked to fly a Boeing 747 but had been given the fuel only for a DC3."[24] UN troops had neither the equipment nor the mandate to compel UNITA or MPLA troops to abide by the peace accord. With the end of the Cold War, the will to fund a large peacekeeping operation to support the peace process in Angola was lacking, and UNAVEM II was given a limited mandate with the assumption that the parties would ensure the success of the peace process, possibly informed by Zartman's hurting stalemate theory of conflict resolution. Like other traditional UN peacekeeping operations, the UN mission mistakenly assumed that the conflict parties would willingly implement the peace accord and not attack civilians and UN forces.

Even if the mission had adequate resources, its restricted rules of engagement to use force only in self-defense excluding the protection of civilians undermined its capacity to alter the calculations of the belligerents to choose peace instead of continued fighting and the killing of civilians. No doubt when civilians were being killed, UN forces looked on helplessly.

In view of the inherent weaknesses of the missions, the conflict parties, especially UNITA, employing the logic of cost–benefit analysis of decision-making, chose continued fighting instead of peace. In the Angolan case, both parties, on the basis of cost–benefit analysis, chose to continue to fight because the Angolan parties were interested in peace only if it came at no cost to their political, military, or economic power. As Anstee observed, "Savimbi was a man with a mission and messianic view of his role in history: he saw himself as the man born to govern Angola and lead the people out of the wilderness,"[25] and therefore anything short of that was a loss to Savimbi. He refused to join the government of Unity and National Reconciliation formed in 1997 because it was headed by his MPLA rival President Neto and dominated by MPLA that had a lion's share of political and economic powers. To protect his political and economic power, Savimbi decided, based on cost–benefit analysis, to renege on his commitments to the peace accords and chose to fight. Of the control variables in the statistical model in Chapter 5, only regime type seems to have contributed to continued fighting and deliberate civilian killings. Angola, during the period of study, was less democratic as evident in reports of electoral malpractices by the incumbent in 1992 that led to renewed conflicts and the killing of civilians. Peace agreements contrary to conventional wisdom did not reduce civilian killings as both parties continuously reneged on their commitments due to mistrust of each other.[26] The other control variables in the statistical model do not seem to have any impact on the number of civilian killings.

Looking at other potential factors excluded from the statistical model, delayed deployment of mission affected deliberate civilian killings. Late approval of the mission's budget as well as stern opposition of the conflicting parties toward deployment of UN forces caused delays. The original plan for UNAVEM II was that all troops were to be assembled by August 1, 1999, but logistical problems did not make it possible. A fundamental problem was lack of food, medical supplies, and clothing and this combined with the caution of commanders and leaders of both conflict parties affected early deployment. The assembly was clearly behind schedule by early 1992.[27]

A serious flaw in the United Nations' effort in Angola was its assumed partiality against UNITA, which questioned United Nations' credibility and legitimacy as an honest broker. UNAVEM II was accused by UNITA of favoring MPLA given United Nations' hostility toward UNITA. UN personnel reportedly were far from being objective and refused to investigate UNITA claims of massive electoral abuses by the MPLA. By this act, the mission was presumed to have compromised its principle of impartiality and undermined its legitimacy as an honest broker. The UN Security Council is also said to have facilitated the MPLA's aggression by passing large numbers of resolutions that singled out UNITA for obstructing the peace process but ignored similar actions by MPLA.[28] UNITA predictably lost confidence and trust in the mission as it viewed the force as a dishonest and biased broker. The mission could therefore not alter UNITA's calculations and raise its value for peaceful negotiations; rather it raised its perceived cost and the cost of involving it in the search for peace. UNITA did not cooperate with the United Nations' efforts to peacefully resolve the conflict and continued to fight leading, for example, to the breakdown of a 16-month ceasefire in late October 1992 and mass killing of vulnerable civilians.

The United Nations' troubled effort in Angola was undermined further by waste and corruption. Tens of millions of dollars were wasted due to procurement fraud involving corrupt UN officials and suppliers, according to an April 28, 1995, report by the UN Under-Secretary for internal oversight.[29]

The covert involvement of the superpowers (the United States and the Soviet Union) in the Angolan war, each supporting a conflict party, affected the capacity of the mission to constitute a formidable barrier to the killing of civilians by the belligerents. The UNITA/MPLA conflict became a Cold War battleground for Washington, D.C. and Moscow with United States/South Africa supporting UNITA and Soviet Union/Cuba/China supporting MPLA government with resources.[30] The support received emboldened and hardened the belligerents' bargaining positions, raising their benefits of choosing conflict rather than peaceful negotiation of the conflict. This situation prolonged the conflict and resulted in more civilian casualties.

Examining mission objectives and outcomes, UNAVEM I was successful in fulfilling its limited mandate of helping oversee the withdrawal of Cuban forces because of the cooperation of the parties. UNAVEM II had three principal tasks and failed consistently at all the three due to its size and resources. It could not verify the demobilization of the forces, could not ensure the neutrality of the police force and monitor adequately the elections. The assistance to monitor the election was very minimal as a mere 89 civil servants were deployed for all electoral activities in the six regional centers and 18 provincial centers throughout the country. UNAVEM II failed and Angola returned to war following the elections in 1992. UNAVEM II had neither the adequate resources nor the authority to help the parties find peace, even if they had been sincere about wanting to do it.

UNAVEM III, which had the resources that UNAVEM II lacked, achieved some peace and helped establish a national unity government but UNAVEM III failed to demobilize the UNITA and fighting resumed. Angola witnessed heavy fighting at the withdrawal of UNAVEM III and transition to MONUA. The United Nations was unable to stop the war or resolve the conflict. UNAVEM III therefore was not successful in accomplishing the mission objectives. MONUA, with a reduced force strength of 2,000 peacekeepers, was asked to accomplish in a few months what 7,000 peacekeepers could not do in over two years during UNAVEM III. MONUA's efforts to deal with the conflict situation were largely unsuccessful as both sides refused to cooperate.

Conclusion

UN traditional peacekeeping in Angola failed because the mission did not constitute a formidable barrier to the killing of civilians by the belligerents in view of its inherent deficiencies as well as difficulties associated with the host country. Apart from their failure to protect civilians, the missions, except UNAVEM I, were not successful in accomplishing their assigned mandates. As a result of inadequate resources, unworkable mandates, restricted rules of engagement, less democracy, lack of cooperation from conflict parties, delayed deployment, partiality of the mission, corruption, and the covert support of conflict parties by major powers, the mission's capacity to alter the calculations of the parties by raising the cost of fighting and benefits of choosing peace was undermined. The belligerents therefore chose the option of continued fighting with attendant killing of civilians in the pursuit of their political goals. Deliberate civilian killings could have been reduced or prevented if the mission had been better designed with appropriate rules of engagement that permit the use of force for self-defense and protect civilians, and equipped with sufficient number of troops and resources to enable the mission to form a formidable barrier to the killing of civilians by altering the calculations of the parties through raised costs of fighting and benefits of choosing peace.

In conclusion, civilian killings could have been reduced if a realistic and well-designed robust mission with a large strength, heavy weapons, and appropriate rules of engagement had been deployed and if the United Nations and the major powers had been honest rather than biased brokers.

UNAMIR: 1993–1996

Rwandan Civil War

Rwanda from 1964 to 1994 had been a theater of massive violence between their main ethnic groups, the Hutu and the Tutsi. The Rwandan civil war and the eventual genocide is the product of longstanding institutionalized discrimination and struggle for socio-politico-economic power along ethnic fault lines between the Hutus and Tutsis. Rwanda typifies how institutionalized discrimination and unequal access to scarce resources lead to violence. Historically, Rwandan society has become polarized along ethnic lines and ethnic exclusion has come to dominate its life. This precarious internal dynamics, reinforced by the Belgian colonizers, culminated in the radicalization of animosity and routinization of violence and the "moral exclusion" of the Tutsis by the Hutus in the post-independence era allowing first their "social death" and eventual genocide in 1994.[31] Precolonial Rwanda had a Tutsi-dominated monarchical system but was profoundly modified by the Belgian colonizers for strategic interests and administrative expediencies. The Belgians turned precolonial socioeconomic stratification into essentialized ethnicity. Political, social, and even economic relations became more rigid, unequal, and biased against the Hutu, while the power of Tutsis greatly increased under the Belgian colonial administration.[32] Tutsis considered themselves "inherently superior" to the Hutu and Twa and exclusively controlled access to all avenues of power in Rwanda.[33] They perceived to have an inalienable right to rule and enjoyed unlimited wealth and prestige from a system of economic relations that cruelly exploited the labor of subordinate Hutu groups.[34] While most Tutsis were wealthy and powerful, almost no Hutus were and most of them suffered untold misery at the hands of the Tutsi "despots."[35] For the marginalized Hutus, the ethnic struggle against an institutionalized, structured Tutsi despotism was as significant as the struggle for independence from the Belgians.[36] Under these conditions, it is not surprising that between 1958 and 1962 a small group of Catholic-educated Hutu counterintelligentsia undertook a revolution to overthrow the monarchy in Rwanda and declared it a Republic in 1961. The Hutu social revolution occurred with the connivance of the departing colonizers, who during the last years before independence in 1962 switched their allegiance to the Hutu out of the fear of the more radical (leftist, anticolonial) Tutsi elite.[37] Under the new Hutu regime, the structural discrimination and rigid social, economic, and political hierarchy that existed under Tutsi rule were reversed favoring the Hutu ruling elite. Habyarimana's regime (1973–1994), for example, institutionalized structures of discrimination against the Tutsis, especially in areas that allowed vertical mobility, such as modern education, state jobs, and politics. Tutsis were denied political equality and strict limitations were placed on their ability to achieve social and economic advancement.[38] The structural inequality profoundly institutionalized by the new Hutu ruling elites further polarized the two groups and laid the foundation for civil war and mass violence in the years to come. The Hutu revolution culminated in three intricate phases of violence and brutality that cost the lives of tens of thousands of Tutsi civilians. In late 1959, there were localized anti-Tutsi violence and small pogroms in some provinces. Hundreds were killed, and many Tutsis fled the country. From 1961 to 1964, some of these Tutsi refugees attempted to return militarily, launching small guerrilla assaults from Burundi and Uganda. The insurgency was foiled, but led to organized mass killings of innocent Tutsi civilians in Rwanda. In early 1962, more than 2,000 Tutsis were killed and at least 10,000 more died in December 1963. During this time, between 140,000 and 250,000 Tutsis fled Rwanda[39] The return of the Tutsi refugees was categorically denied with the argument that there was no more space in Rwanda.[40]

The Rwandan war took a dramatic turn in 1986 when the Ugandan rebel army, which included many Rwandan Tutsi refugees, overthrew the government of Uganda. In the light of continuing discrimination within Uganda, the Rwandan contingent of the Ugandan rebels formed the Rwandan Patriotic Army (RPA) and a political counterpart, the Rwandan Patriotic Front (RPF), and invaded Rwanda in October 1990, displacing hundreds of thousands of Rwandans internally. The conflict attracted foreign actors' involvement with Uganda supporting the RPA while Zaire, Belgium, and France openly provided military support to the Rwandan Hutu dominated government. An initial Zairian mediation effort achieved a ceasefire in March 1991, but Habyarimana immediately renounced it because of alleged RPF violations.

The Rwandan government had little incentive to compromise at the time, since it had gained the upper hand in the war.[41] International aid donors further pressurized the government and trading partners utilized their leverage to insist on reform. Then, in late 1991, the RPA resumed their attacks under the leadership of Commander Paul Kagame.[42] Confronted with these dynamics, President Habyarimana initiated nominal steps toward political pluralization, and in 1991 he legalized multiple political parties. Simultaneously, however, Habyarimana established two new institutions to empower extremists and entrench his authority: the CDR (a right-wing political party) and the *Interhamwe* militia (a paramilitary organization that was to play a major role in the genocide). Then, in April 1992, Habyarimana announced a multiparty cabinet but it fell short of democratic practices since no elections were held and the ultimate authority resided in the presidency.

The RPA resumed its insurgency with successes in occupying parts of Rwanda. In view of RPA's success, Habyarimana acknowledged the need for a peaceful resolution and therefore agreed to the mediation efforts by concerned international actors to reach an agreement in Arusha, Tanzania, in July 1992. The negotiation was brokered by the host nation with the support of all key regional and international interests: France, Belgium, the United States, Uganda, Zaire, Burundi, the United Nations, and the Organization of African Unity (OAU).

In October, however, Habyarimana backed away from the negotiating process[43] and the RPA resumed its insurgency gaining some territories and displacing millions of civilians, forcing Habyarimana to the peace table again.[44] Finally, on August 4, 1993, the contending parties reached a historic peace agreement at Arusha, called the Arusha Accord, laying out a comprehensive roadmap for a transition to genuine multiparty democracy. The Arusha Accords mandated the creation of a Broad Based Transitional Government (BBTG) and a Transitional National Assembly (TNA) wherein power was to be divided between the RPF and the major political parties in Rwanda.[45] The agreement also reduced the powers of the president and made the military an integrated force to be comprised of soldiers and officers from the Rwandan army, the Forces Armées Rwandaise (FAR), and the RPF.[46] However, the Hutu elites and extremists, including the former ruling party, the Mouvement Républicain National pour la Démocratie et le Développment (MRNDD) and its extremist offshoot, the Coalition pour la Défense de la République (CDR), perceived the Arusha Accord as completely illegitimate.[47]

The general consensus among these powerful segments of the Hutu political elite was that sharing political power with the Tutsis in a multiparty system of government was an anathema to Hutu interest. The Accord was perceived to have substantially reduced the power of the MRNDD relative to the opposition, particularly in terms of the disproportionate amount of power it ceded to the RPF. The number of positions held by the RPF in the BBTG, the TNA, and the armed forces were disproportionally higher than Hutus comparing their relative total percentage of the total Rwandese population.[48] President Habyarimana and moderate

Hutus were blamed for "selling out" to the RPF. Their displeasure with the process of political reform culminated in a strategy to liquidate the moderate Hutu opposition and the entire Tutsi population in the spring of 1994. Tutsis became the subject of a hateful propaganda that included explicit and regular incitation to mass murder, verbal attacks, the publication of lists with names of people to be killed, and threats to anyone having relations with Tutsi. Between 1990 and 1993, thousands of Tutsis were massacred by mobs directed by local authorities, national politicians, and police.[49] Large amounts of arms including machetes were imported and distributed to the militia for action on the "D-day." The extremists' radical rhetoric and routinized violence further dehumanized the Tutsi and legitimized violence.[50] On April 4, 1994, a plane carrying Rwandan president Habyarimana and Burundi president Ntaryamira was shot down, killing both presidents. Their death fuelled extreme violence and instability by Hutu *génocidaires*, massacring about 800,000 Tutsis and moderate Hutus between April 6, 1994 and July 18, 1994. The RPF resumed the civil war and conquered Kigali in July itself, which signaled the end of the genocide.

UNAMIR deployed

During the Arusha peace agreement in August 1993, both parties requested the United Nations to assist in the implementation of the agreement. Subsequently, the Security Council by its resolution 872 (1993) established the UNAMIR with a mandate to assist in ensuring the security of the capital city of Kigali; monitor the ceasefire agreement procedures, including establishment of an expanded demilitarized zone and demobilization procedures; monitor the security situation during the final period of the transitional government's mandate leading up to elections; assist with mine-clearance; and assist in the coordination of humanitarian assistance activities in conjunction with relief operations.[51] An understrength UNAMIR mission of 2,548 lightly armed troops, with limited rules of engagement to use force only in self-defense, was deployed in 1993.

After Rwandan president Habyarimana and Burundi president Ntaryamira were killed on April 4, 1994, and the genocidal killings started, UNAMIR Commander Romeo Dallaire requested for more troops and greater authority to stop the killings of civilians. With massacres continuing on a large scale, the Secretary-General urged the Security Council to take a more forceful action to restore law and order, though to do so would require a much greater commitment than Member States appeared willing to consider. The Secretary-General suggested to the Security Council three options: to massively reinforce UNAMIR; to reduce its size to a small group to remain in the capital acting as an intermediary to secure ceasefire; or to withdraw altogether. Under the Security Council resolution 912 of April 21, 1994, the second option was taken. UN officials rejected the force commander's request and ironically reduced the troop size from 2,548 to 270 with orders to use force only in self-defense in the midst of these atrocities and within a month almost 800,000 to a million civilians were killed while UN personnel helplessly looked on. General Dallaire, despite orders to withdraw from Kigali, refused to abandon the country to the genocide, and remained to help. An understaffed and abandoned UNAMIR did the best it could to save the lives of thousands of Tutsis in and around Kigali and the few areas of UN control. With a reduced force and restricted rules of engagement, the United Nations lacked the appropriate capacity to successfully negotiate a ceasefire and bring an end to the killings. As a result of the mission's failures, the Secretary-General pushed for increased force strength in April 1994 and after delays, the Security Council by its resolution 918 (1994) on May 17, 1994, beefed up UNAMIR II force of 5,000 troops but was not permitted to deploy until July 1994 after the genocide had ended. UNAMIR and UN officials were criticized for not stopping the genocide. Following the end of the full-scale war, a new government of national unity was established on July 19, 1994. The new government questioned the role of UNAMIR with a reduced strength of 1,800 forces and demanded the end of the mission, which finally happened on March 8, 1996.

Analyses

UNAMIR, like UN missions in Angola, failed to stave off intentional civilian killings by belligerents because its traditional designation rendered it incapable of constituting a formidable barrier to alter the belligerents' calculation to choose peaceful negotiations rather than war. UNAMIR was guided by traditional peacekeeping principles of small arms, impartiality, with rules of engagement authorizing the use of force only in self-defense. UNAMIR was also under-strength with a force strength of 2,548 without major power participation. UNAMIR's capacity to raise the costs of continued fighting and benefits of choosing peace for belligerents was crippled by its lack of resources and appropriate mandate and rules of engagement to use force not only in self-defense but also to protect civilians under imminent threat. The lightly armed, understrength and ill-equipped UNAMIR, with restricted rules of engagement that permitted the use of force only in self-defense excluding the protection of civilians, could neither deter nor stop the killing of about 500,000 civilians in 1994.

In the case of Rwanda, all the variables that could have lowered intentional civilian killings were absent. First, as recognized by the conflict parties, the designation of traditional peacekeeping rather than "robust" peacekeeping was the first flaw leading to mission failure in preventing intentional civilian killings. Both conflict parties, during the Arusha Accord, recognized that successful implementation would require a secure environment in view of the passionate stakes involved in ethnic conflicts. Article 54 of the peace accord accordingly called for a neutral international force to provide security for the process and civilians and the United Nations offered to do so.[52] The concerns of the conflict parties and subsequent UN guarantee to provide secure environment clearly implied that a mission force more capable than a traditional one was needed. UNAMIR therefore required a robust mission with greater capabilities to constitute a formidable barrier to belligerents' choice of continued conflict and attendant civilian killings and alter their calculations to choose peaceful resolution of the conflict. Facing a robust mission with a large force, adequate resources, and robust rules of engagement to use force to protect civilians, use of heavy weapons, and major power participation, Hutu extremists would have seen the mission as a formidable obstruction and the fear of incurring costs both human and otherwise would have made them choose peace. UNAMIR, however, lacked these capabilities and resultantly failed to prevent the genocide. The mission's vague mandate, created under Chapter VI of the UN Charter was unclear about the right to use force, particularly in defense of civilians and stopping the genocide. The circumstance leading to the murder of the ten Belgian troops typifies the confusion over legal protocols on the use of force in UNAMIR. The Belgian troops of second Commando Battalion, the Paracommando Regiment, were killed after they were advised by their battalion commander to hand over their weapons to Rwandan government troops because the Commander was unclear on the legal issues in respect of the use of force for self- defense despite being under fire for approximately two hours. Security Council resolutions for UNAMIR mission were vague and unclear on the right to use force in stopping the genocide. In one of Romeo Dallaire's parting cables, he said that "the [UN] force has been prevented from having a modicum of self-respect and effectiveness on the ground."[53]

UNAMIR did not have the requisite resources to perform its mandate.[54] The United Nations had difficulty finding peacekeepers and funds for the mission. UNAMIR's budget was not approved until April 4, 1994, two days before genocide broke out. As a result, UNAMIR operated on a limited budget and lacked essential supplies, equipment, and vehicles. In fact funding was so limited that between October 1993 and April

1994, UNAMIR contingents at one stage borrowed from UNICEF simply to remain operational.[55] Worse still, the United Nations deployed lightly armed 2,548 peacekeepers, less than half the recommended contingent with rules of engagement restricting the use of force only to self-defense and excluding the protection of civilians under threat as promised in the peace accord. Following the death of the two presidents of Rwanda and Burundi, UNAMIR Commander Romeo Dallaire requested for more troops and greater authority to stop the killings of civilians. UN officials rejected his request and ironically reduced the troop size from 2,548 to 270 with orders to use force only in self-defense and within a month almost 800,000 to a million civilians were killed whiles UN personnel helplessly looked on. Evidently, Major General Romeo Dallaire had strong convictions that increased strength with robust rules of engagement could have deterred or stopped the genocide when he stated that if he "had had a brigade group in Rwanda, there would be hundreds of thousands of lives spared today."[56] Unfortunately, many UN Member States delayed contributing personnel for some time, until the main wave of killings ceased.

UNAMIR also suffered from the lack of major-power interests, existence of which could have made the major powers to step in and stop the violence. The international community was instrumental in the signing of the accord that prescribed a robust mission, but due to major powers, especially the United States, disinterest in the conflict, Rwanda was handed a poorly designed traditional mission that could neither protect itself nor protect the defenseless civilians. Coincidently, the establishment of UNAMIR on October 5, 1993, followed two days after 18 Americans troops and 500 Somalis were killed during a mission in Somalia. Following the Somalia debacle, the interest of the United States and other major powers to support the robust mission envisioned for UNAMIR in the Arusha agreement waned. It was a time the United States was attempting to extricate itself from Somalia and limit its role as the "world's policeman."[57] President Clinton clarified United States' position with a statement in his commencement address at the US Naval Academy in 1994: "we cannot solve every outburst of civil strife or militant nationalism simply by sending in our force."[58]

In view of the inherent weaknesses of UNAMIR, extremists Hutus calculated that sharing power with the Tutsis would cost their desire for absolute political and economic domination and therefore chose to exterminate the Tutsis and moderate Hutus who supported peace settlement through power sharing.

Examining the control variables in the statistical model in Chapter 5, the ethnic nature of the conflict greatly affected deliberate civilian killings.

Historic animosity between the Hutus and the Tutsis dating back to the colonial era that led to the earlier conflict and civilian killings of 1964 was the underlying cause of the 1994 genocide. Extremist Hutus were so passionate about exterminating the Tutsis that even moderate Hutus who wanted peaceful conflict resolution were killed.

Rwanda was less democratic without consolidated democratic institutions to promote peaceful conflict relation and the nominal steps taken toward political pluralization were flawed by the concentration of power in the presidency. The politics of exclusion by the Hutus and the Tutsis' pursuit of political inclusion fuelled the conflict resulting in the genocide. Other factors such as borders, peace agreement, cultural similarity, and the cultural composition of the mission force did not seem to have affected the deliberate killing of civilians.

Examining other potential causal factors excluded from the statistical model in Chapter 5, we find that delays in the deployment of UNAMIR force also impacted on UNAMIR's capacity to stave off intentional civilian killings. During the accords, the conflict parties appealed to the United Nations to quickly deploy troops to enhance the implementation of the accord, which the United Nations promised to do. Both parties recognized that delay in implementation might enhance the intentions of Hutu extremists to pursue their genocidal agenda.[59] The United Nations, contrary to its initial promise, delayed for over two months after the peace agreement before authorizing the deployment of UNAMIR forces. In fact, most troops arrived several months later, some only a couple of weeks before the genocide began.[60] Apart from its failure to stave off civilian killings, UNAMIR failed to accomplish its mandate due to its inherent deficiencies previously discussed.

Even prior to the genocide, the UN Secretary-General conceded that there were insufficient civilian police monitors to fulfill one part of the mandate, "to investigate and report on incidents regarding the activities of the gendarmerie or police."[61] Yet, it is hard to see how increased investigation and reporting would have prevented the eventual genocide. Even a several-fold increase in the number of UN peacekeepers would not have secured peace, so long as the restrictive Chapter VI mandate remained in effect. Major-power disinterest in Rwanda also accounted for mission failure.

In the following months after the end of the genocide, UNAMIR continued its efforts to ensure security and stability, support humanitarian assistance, clear landmines, and help refugees but the new Tutsi-led Rwandan government supported ending the mission, stating that UNAMIR did not respond to its priority needs. The Security Council heeded that request, and UNAMIR left in March 1996.

Conclusion

UNAMIR was deployed to keep peace in troubled Rwanda, yet was given neither the appropriate mandate nor the resources to accomplish its mission due to major powers' disinterest in the mission. UNAMIR peacekeepers therefore could not constitute a formidable barrier enough to deter extremists or halt the genocide as a result of its inherent deficiencies. The traditionally designed UNAMIR with limited resources (personnel, equipment, funds etc.), restricted rules of engagement to use force only in self-defense, and disinterest of major powers in the mission adversely affected the mission's capacity to alter the calculations of the belligerents to choose peace over continued fighting with attendant civilian killings in the pursuit of their political goals. The Rwandan tragedy was a tragic result of using inappropriate traditional mechanism, fed by the disinterest of the major powers in Rwanda, to resolve a civil war bitterly caused by passionate ethnic hatred. The killing of civilians was also impacted by the ethnic nature of the conflict, less democratic political structures and delayed deployment of the mission force. Had the United Nations and global powers faithfully designed a timely deployed and well-resourced robust mission in accordance with the terms of the Arusha accords, the mission could have formed a formidable barrier to stave off civilian killings by the belligerents as well as accomplishing its mandate.

The United Nations Mission in Sierra Leone (UNAMSIL): 1999–2005

Sierra Leonean Civil War

Sierra Leone was wracked by a devastating conflict (1991–2002) characterized by gross human rights violations with over 50,000 dead.[62] The Sierra Leonean civil war was the product of colonialism and decades of post-independence corruption among the political elite.[63] With the support of Liberian rebel leader Charles Taylor, Foday Sankoh's Revolutionary United Front (RUF) entered south-eastern Sierra Leone in March 1991. The RUF claimed to have reignited radical pan-African revolution in Sierra Leone by acts of protracted insurgencies against incumbent governments that resulted in the killing of tens of thousands and the displacement of millions of people. Despite RUF's populist rhetoric of emancipating the populace from perceived exploitation and corruption by the government's elites, the impoverished rural population rather than the elite were targets of RUF's campaign of terror.[64] By 1992, the RUF was in control of the diamond-rich Kono region and proceeds from the diamond sales financed and sustained its campaign and enhanced its control over some parts of Sierra Leone.[65] President Momoh's inability to pay government troops provoked a military coup led by Captain Valentine Strasser in 1992.[66] The National Provisional Ruling Council (NPRC) government used various strategies including the hiring of Executive Outcomes, a South African mercenary company,[67] to retake some of the diamond mines.[68]

Despite another coup by Julius Maada Bio,[69] local and international pressures led to February 1996 presidential elections won by former UN official Tejan Ahmad Kabbah. Pressure from regional and international stakeholders and on the brink of defeat by the Civil Defense Forces (CDF), the RUF acquiesced in peace negotiations. The resulting Abidjan Agreement of November 1996, however, collapsed within a year.[70] In 1997, the Kabbah administration was overthrown by the Koromah-led Armed Forces Revolutionary Council (AFRC), which invited the RUF rebels to join the coalition government. Rule of law collapsed over subsequent months amidst serious human rights violations. The situation provoked UN and ECOWAS sanctions on the regime compelling the AFRC/RUF government coalition to the negotiating table in Conakry, Guinea.[71] Under international and local pressure during the Conakry Accords, the AFRC/RUF agreed to demobilize and return Kabbah to power in October 1997 but it reneged. Nigerian forces, acting under regional mandate, finally ousted AFRC on February 1998, to reinstate Kabbah, forcing RUF to retreat to guerrilla war tactics backed by significant numbers of Sierra Leone soldiers living off land and financed by "blood diamond" trade. Despite the Nigerian intervention, the RUF/AFRC continued to fight voraciously in the countryside and the conflict reached a stalemate. The presence of ECOMOG forces weakened RUF and Sankoh's position who had been captured in Nigeria at the time of the breakdown of the Abidjan Agreement and sentenced to death following Kabbah's reinstatement in 1998,[72] paving the way for renewed peace efforts. In July 1998, the Security Council set up the UN Observer Mission in Sierra Leone (UNOMSIL) to monitor the peace and disarmament process.[73] The international community could, however, not prevent the rebels from launching another offensive in December 1998. The rebel attack, "Operation No Living Thing," on Freetown in January 1999 was the most atrocious chapter of the war, which considerably weakened Kabbah's government.[74] The rebels seized parts of Freetown, but were eventually repelled back by ECOMOG.

A military stalemate became discernible in May 1999 and the major actors sponsoring ECOMOG's operation, particularly Nigeria, were unwilling or unable to spend additional resources to defeat the RUF.

Meanwhile, the United Nations was unwilling to extend its involvement in the situation with no plans to deploy a Chapter VII robust mission in Sierra Leone.[75] These two factors conspired to pave the way for negotiations in Lomé, Togo. The July 1999 Lomé Agreement was a power-sharing and amnesty agreement between the government and RUF intended to end the Sierra Leonean civil war between the RUF rebel alliances and the government of Sierra Leone.[76] It first granted blanket amnesty to RUF rebels and provided for a new government of national unity headed by Kabbah with Sankoh as vice-president. The Lomé Peace Accord also provided for the establishment of the Truth and Reconciliation Commission mandated "to address impunity, break the cycle of violence, provide forum for both the victims and perpetrators of human rights violations to tell their story and get a clear picture of the past in order to facilitate genuine healing and reconciliation."[77] Presidential and parliamentary elections were envisaged for 2002.

In October 1999, the UN Security Council by its resolution 1270 (1999) of October 22, 1999, established UNAMSIL to assist with the implementation of the Lomé Peace Accord. However, Sankoh reneged on the agreement and used his position to engage in illicit trade in diamonds[78] and attack UNAMSIL. Following continued human rights violations and instability, the government of Sierra Leone wrote to the UN Secretary-General requesting the establishment of a Special Court to bring RUF leaders to justice.[79] In August 2000, UN Security Council under UNSCR resolution 1289 (2000), in response global pressure, handed UNAMSIL a robust mandate to robustly counter RUF attacks,[80] especially after the withdrawal of ECOMOG starting December 1999.[81] The combined efforts of UNAMSIL, the Sierra Leone Army, and a British intervention successfully defeated the RUF offensive.[82] Foday Sankoh was arrested during the offensive and paved the way for the conclusion of the Abuja Cease-Fire Agreement.[83] The UN Secretary-General authorized under Security Council resolution 1315 the establishment of a Special Court for Sierra Leone. On January 16, 2002, the United Nations and the Sierra Leonean government signed an agreement establishing the Special Court.[84] The Special Court would prosecute "persons . . . bearing the greatest responsibility for the commission of crimes against humanity, war crimes and other serious violations of international humanitarian law, as well as crimes under relevant Sierra Leone law committed within the territory of Sierra Leone."[85] The strong presence of UNAMSIL troops and a newly constituted, British-trained Sierra Leonean army enhanced peace[86] and with the completion of the disarmament program, an official declaration ending the war followed shortly after the signing of the Special Court Agreement.[87]

UNAMSIL deployed

The Lomé Agreements demanded the deployment of a neutral peacekeeping force in Sierra Leone to assist with the implementation of the agreements. The UN Security Council by its resolution 1270 (1999) of October 22, 1999, established UNAMSIL to assist with the implementation of the Lomé Peace Accord. UN Security Council established 6,000-strong UNAMSIL force (eventually increased to 17,500) under Chapter VI of UN Charter, to replace departing ECOWAS troops. UNAMSIL at its inception was designed in traditional terms, which was reflected in the deployment of lightly armed weak force strength of 6,000 personnel authorized to use force only in self-defense and without major power participation. The available troops' strength was too low to allow a widespread deployment and deter RUF attacks on civilians and the peacekeepers. UN peacekeepers were often denied freedom of movement amidst violations of ceasefire, including attacks against civilians and UNAMSIL peacekeepers by the RUF.[88] A bipolar situation existed with the RUF controlling the north and east of the country with the government of Sierra Leone or UNAMSIL controlling the west and south.[89]

Realizing UNAMSIL's incapacitation due to its poor strength, inadequate resources, and its defensive posture, the RUF attacked the peacekeepers and civilians culminating in widespread killings of civilians and hostage-taking of 500 UN peacekeepers in May 2000. Following these attacks, the entire peace process reached an impasse and UNAMSIL realized that successful implementation of mission mandate would require an effective deterrence capacity. The situation worsened with the withdrawal of the regional forces (ECOMOG) and the increasing rebel violations made the Secretary-General to demand for a sizable and credible UN force capable to deter the RUF from derailing the peace process. Upon the Secretary-General's recommendation, the UN Security Council under UNSCR resolution 1289 (2000) increased UNAMSIL's strength to 11,000 personnel and handed it a robust mission to ensure the security and freedom of movement of UN and humanitarian personnel, and protect civilians at risk of attack. UNSC resolution 1299 (2000) on May 19, 2000, further increased UNAMSIL's force strength to 13,000 military personnel.[90] UNAMSIL was mandated among others "to take the necessary action to ensure the security and freedom of movement of its personnel and to afford protection to civilians under imminent threat of violence . . . ."[91]

Despite the fact that the resolution authorized UNAMSIL to intervene and protect civilians under imminent threat, UNAMSIL still lacked the capacity to act in view of its poor state of equipment and training of personnel. The UN Secretary-General therefore accordingly called upon major powers to deploy highly trained and well-equipped combat-ready forces in Sierra Leone to stabilize the precarious situation and bolster the UNAMSIL force. Secretary-General Annan consulted with the United States and United Kingdom but both refused. The United Kingdom, however, later on provided a battalion of paratroopers and five warships giving UNAMSIL "muscle" to stabilize the situation and protected civilians.[92] By November 2001, UNAMSIL reached its authorized strength of 17,500 and was better equipped than at any time since its inception. The Pakistani Brigade, the Russian air wing, and the British Forces (BRIFOR) together provided the mission with powerful deterrent capacity to deal firmly with the RUF—spoilers of the peace process. With enhanced capacity through increased strength, equipped with heavy weapons, major power participation and rules of engagement that authorized the use of force to intervene and deter spoilers, UNAMSIL succeeded in deploying in all districts of Sierra Leone and afforded a degree of protection, freedom of movement, and delivery of humanitarian assistance to civilians. UNAMSIL conducted cordon-and-search operations against the RUF rebels. In addition to protecting civilians from rebel attacks, between May 18, 2001, and January 6, 2002, 45,449 former combatants handed their weapons to the authorities.[93] End of Sierra Leone's 11-year civil war was proclaimed in January 2002 and Kabbah was re-elected president in May 2002. UNAMIR concluded its mandate and withdrew at the end of December 2005.[94] United Nations Integrated Office for Sierra Leone (UNIOSIL), deployed to help maintain stability in the country, succeeded it.

Analyses

The Sierra Leonean case supports the core hypothesis that robust peacekeeping works better than traditional peacekeeping in lowering intentional civilian killings. It also supports the *formidable barrier* model argument that a UN peacekeeping mission's capacity to stave off civilian killings must be able to alter the calculations of belligerents to choose peace instead of conflict by raising the costs of continued conflicts and benefits for the belligerents. The mission's capacity to achieve that depended on, as hypothesized in Chapter 3, large force size, use of heavy weapons, authorization to use force for self-defense as well as to protect civilians under eminent threat, and major power participation in the mission. UNAMSIL, which was initially designed in traditional terms, faulted by deploying a poorly equipped weak force with the rules of engagement restricted to using force only in self-defense and without well-trained, well-supported major power participation. UNAMSIL in its early stages could not constitute a formidable barrier to the rebels' killing of civilians because its small force size, restricted rules of engagement, use of small weapons, and absence of major power contingents in the mission were not deterrent enough to alter the calculations of the rebels to choosing the option of peace over continued conflict with attendant civilian killings in the pursuit of their political goals.

UNAMSIL, as a result of its initial weaknesses, could not raise the costs of killing civilians and benefits of choosing peace for the rebels, and therefore it failed to stave off civilian killings in Sierra Leone. The rebels realized that the small-sized UNAMSIL force with light weapons, without major power participation, and authorized to use force only in self-defense were not formidable enough to impose on them any costs (human or otherwise). In view of this, the choice of continued fighting with attendant civilian killings looked the better option since that gave them absolute control over the diamonds in their occupied territories. In addition, the choice of continued fighting enhanced their bargaining position, forcing the government to make concessions to them, which was what happened when the Lomé Accord granted them blanket amnesty and made the rebel leader Foday Sankoh the vice-president. In line with the formidable barrier of robust peacekeeping model, UNAMSIL realized that successful implementation of mission mandate would require an effective deterrence capacity to alter the calculations of the rebels to choose peace rather than continued killing of civilians. In the face of its disastrous failures reflective in mass civilian killings and attacks on UNAMIR forces, UNAMSIL eventually changed its operational paradigm to a robust mission.

The UN Security Council under UNSCR resolution 1289 (2000) increased UNAMSIL's strength to 11,000 personnel (17,500 at its maximum strength) and handed it a robust mission to ensure the security and freedom of movement of UN and humanitarian personnel, and protect civilians at risk of attack. In addition to its increased strength and robust rules of engagement, the deployment of the heavily equipped British battalion together with the Russian air wing provided the mission with a powerful deterrent capacity to deal firmly with the RUF—the spoilers of the peace process.

The strengthened and adequately equipped UNAMSIL force with major powers support and robust rules of engagement succeeded in deploying in all districts of Sierra Leone and afforded a degree of protection, freedom of movement, and delivery of humanitarian assistance to civilians. UNAMSIL conducted cordon-and-search operations against the RUF rebels. UNAMSIL, with enhanced ability, was able to alter the calculations of the rebels to choose the option of peace because the UNAMSIL's size and offensive capacity raised the costs of fighting and killing civilians for the rebels. The rebels incurred human costs (including the death of their leader Foday Sankoh) as well as revenues from the occupied diamond lands during UNAMSIL attacks on their positions. As a result of the rising costs of continued fighting and killing of civilians, the rebels chose the option of peace and between May 18, 2001 and January 6, 2002, 45,449 former combatants handed their weapons to the authorities. The option of peace as a result of rising cost of fighting eventually reduced intentional civilian killings and finally led to peace in Sierra Leone.

Looking at the effects of the control variables in the statistical model, we find that with the exception of borders, none of the factors affected the number of civilian killings. Sierra Leone shares borders with Liberia and Guinea and with limited number of borders one expected that borders would not impact the conflict. On the contrary, the Sierra Leonean conflict was fuelled by Charles Taylor's interest in the conflict and his support of the RUF rebels. His support strengthened and hardened the RUF rebels to continue to fight and kill civilians.

Examining UNAMSIL's mandate, UNAMSIL has been described as a success story or a model of peacekeeping.[95] UNAMSIL completed most of the tasks by the Security Council. Apart from protecting the civilians, the mission disarmed tens of thousands of ex-fighters, assisted in holding national elections, helped to rebuild the country's police and contributed toward rehabilitating the infrastructure, stopped illicit trading in diamonds, assisted the voluntary return of more than half a million refugees and internally displaced persons, monitored and trained Sierra Leoneans in human rights, and assisted in setting up the Special Court of Sierra Leone to try those most responsible for war crimes and a Truth and Reconciliation Commission.

Conclusion

In the Sierra Leone mission, a more proactive style of peacekeeping became a model for successful peacekeeping for providing security for the people through the deployment of robust peacekeeping mission that posed a formidable barrier to the rebels' capacity to kill civilians. The enhanced robust UNAMSIL mission was able to alter the calculations of the rebels by raising the costs of fighting and benefits of choosing peace leading to reduced intentional civilian killings. The Sierra Leonean case therefore supports the *formidable barrier* model as a viable peacekeeping model for mission success. It also supports my core hypothesis that robust peacekeeping works better than traditional peacekeeping in lowering intentional civilian killings. The case also supports the arguments that large troop size, use of heavy weapons, rules of engagement that permit use of force for self-dense and to protect civilians, and major power participation in missions enhance the missions' capacity to protect civilians from intentional killings by belligerents in civil conflicts.

United Nations Organization Mission in the Democratic Republic of Congo (MONUC)/United Nations Organization Stabilization Mission in the Democratic Republic of the Congo (MONUSCO): 1999–present

Democratic Republic of Congo War

Since 1994, the DRC has been wrought by ethnic strife and civil war, referred to as "Africa's first World War," involving seven neighboring nations resulting in the death of millions of people.[96] The war started as a revolution in Zaire by rebel leader Laurent-Désiré Kabila that replaced a decades-long dictator, Mobutu Sésé Seko. Prior to this there were centuries-long interethnic tensions and animosities between various ethnic groups in eastern Zaire, especially between the agrarian tribes native to Zaire and semi-nomadic Tutsi tribes that had emigrated from Rwanda at various times.[97] The historical ethnic tensions there precipitated into war following the 1994 Rwanda genocide and the establishment of a Tutsi-dominated government that sparked emigration of some 1.2 million Rwandese Hutus including the Hutu *génocidaires,* fearing retaliation, to eastern DRC (formerly Zaire),[98] an area inhabited by ethnic Tutsis. Mobutu, rather than controlling the former *génocidaires*, actually supported them with training and resources for an invasion of Rwanda,[99] compelling Rwanda to respond to the security threat posed by the *génocidaires.* Destabilization in eastern Zaire resulting from exacerbated ethnic tensions and antagonisms offered an opportunity to the Kabila-led Alliance of Democratic Forces for the Liberation of Congo (AFDL) to rebel against Mobutu in 1996. This facilitated both internal and external stakeholders to align against and topple the corrupt and inept Mobutu government. The new Kabila government renamed Zaire the Democratic Republic of Congo but failed to address the underlying causes of the war. Interethnic animosities in eastern DRC, where the government lacked control, exacerbated and the economy deteriorated further under Kabila's corrupt rule.[100] Since Kabila gained power with the help of foreign actors, Congolese saw him as stooge of the imperialists who have neorealist economic interests in DRC's mineral resources. Kabila, in demonstrating his political independence to boost his domestic image and support, alienated his foreign allies by expelling all foreign forces from the DRC on July 26, 1998. Kabila's vigorous centralization campaign renewed conflict with minority groups in the east, who demanded autonomy. In August 1998, two brigades of the new Congolese army, with the support of Rwanda and Uganda, rebelled against Kabila. Within weeks, the rebels, Congolese Rally for Democracy (RCD), seized large areas of the country.

Through Security Council diplomatic interventions, the DRC, along with Angola, Namibia, Zimbabwe, Rwanda, and Uganda, signed the Lusaka Ceasefire Agreement on July 10, 1999, for a cessation of hostilities, but the RCD failed to sign. The Lusaka agreement asked all conflict parties to cooperate, under a Joint Military Commission, in tracking, disarming, and documenting all armed groups in the Congo. Subsequent to the signing of the Lusaka Ceasefire Agreement, the Security Council established and deployed MONUC in 2000 by Security Council resolution 1291 (2000) of February 25, 2000. With a force strength of 5,537 troops, MONUC was mandated to support the ceasefire and assist in the implementation of the Lusaka Ceasefire Agreement. However, fighting continued between rebels and government forces throughout the country and between Rwandan and Ugandan forces. On January 16, 2001, President Laurent Kabila was shot and wounded by a bodyguard and died two days later. His son, Joseph Kabila, was sworn in as president to replace him by unanimous vote of the Congolese parliament.

With the deployment of MONUC, peace negotiations continued throughout 2002 to work out the withdrawal of Rwandan and Ugandan troops in exchange for the demobilization and disarmament of Rwandan Hutu rebels and Uganda rebel group, Lord's Resistance Army (LRA), in the DRC territory. In December 2002, all Congolese belligerents and political groups signed a peace deal in Sun City, South Africa. The Pretoria peace deal stipulated that all foreign belligerents would withdraw their armies from the Congo, and the Kabila government would disarm the Rwandan Hutu and Ugandan rebel fighters on its part of the Congolese territory. Under heavy international pressure, especially from the United States, Uganda and Rwanda pulled out most of their troops. By the end of 2002, Rwanda and Uganda claimed full withdrawal from the DRC, although proxies remained. Following this, the Global and All-Inclusive Agreement was signed on December 17, 2002, by Congolese parties of the Inter Congolese Dialogue, namely, the national government, the MLC, the RCD, the RCD-ML, the RCD-N, the domestic political opposition, representatives of civil society, and the Mai Mai. The Agreement obliged the parties to enhance reunification of the country, disarm and integrate the warring parties, and hold elections. The peace deal ushered in a transitional government in June 2003 in which President Kabila shared power with four vice-presidents. However, the reluctance of the former warring parties to give up power to a centralized and neutral national administration weakened the Transitional Government. DRC's first democratic elections in 46 years were held on July 30, 2006, after the popular approval of a new constitution.

The withdrawal of most Rwandan and Ugandan troops in 2002 has not fundamentally changed the dynamics of the conflict. Despite significant demobilization in 2005, particularly in the Ituri province, serious fighting continues in eastern Congo, particularly Kivu and Ituri Provinces, which have been the main theatres for direct and proxy confrontation between local, national, and regional participants in the Congolese conflict since the ceasefire was signed in Lusaka in 1999. The population there is suffering enormously.[101] After successful democratic elections in 2006, violence has re-erupted in the eastern region of the DRC and security conditions continue to deteriorate, particularly in the volatile provinces of Kivu and Ituri. Violent confrontations between the Congolese military (FARDC) and the armed rebels, especially the group led by dissident General Laurent Nkunda, claiming to fight on behalf of the Tutsi minority in the east, have precipitated new waves of internal displacement and have caused a sharpening of ethnic tensions throughout the region. Foreign and internal armed groups perpetrating violent crimes and intimidating civilians throughout North and South Kivu include the forces of Laurent Nkunda, the Forces Démocratiques de Libérations du Rwanda (FDLR) (forces led by the remnants of the Rwandan *genocidaires* who fled to DRC—then Zaire—in 1994) and the Mai Mai militia.

Deployment of MONUC/MONUSCO

Following the Lusaka Ceasefire Agreement, the Security Council established and deployed United Nations Organization Mission in the Democratic Republic of Congo (MONUC) by its resolution 1279 of November 30, 1999, to restore peace. MONUC was established as a traditional Chapter VI peacekeeping mission to observe and monitor the disengagement of forces and maintain liaison with all parties to the Ceasefire Agreement with the deployment of only 500 military observers (MILOBs). Due to the violent conflict environment in which the peace mission was operating, MONUC quickly evolved toward a more robust Chapter VII operation when the Security Council by its resolution 1291 (2000) of February 25, 2000, authorized MONUC with a force strength of 5,537 troops to take "all the necessary means to protect civilians under imminent threat of physical violence"[102] including the use of force. MONUC is designed in robust terms. It is well equipped with heavy weapons such as attack helicopters, with force strength of 17,000 at its peak, and it is also the most financed UN operation with an allocation of US$1 billion per year.[103] However, the underequipped mission has produced mixed results. The Kisangani massacre of more than 180 people near a UN camp in May 2002 illustrates the gap between MONUC's mandate and the capacity to support it.[104] However, it protected 11,000 civilians who fled to the Bunia airport camp during tribal conflicts in the east Ituri district. Following MONUC's incapacitation to protect the civilians, the European Union (EU) and the UN deployed the International Emergency Multinational Force (IEMF) Artemis14 to Bunia for three months under UN Security Council resolution 1484 (May 2003). In September 2003, the European Union force (EUFOR) led by Germany and France conducted in Ituri "Operation Artemis" to stabilize the east capital of Bunia in September 2003. The well-trained and well-equipped German and French forces, deployed under Chapter VII to enforce peace, successfully rescued the peace process by restoring peace there. Again, from July to December 2006, EUFOR helped MONUC to maintain stability, especially during the elections.[105] The Security Council, by its resolution 1493 (July 2003), authorized the increase of MONUC's military strength to 10,800 personnel. However, MONUC's mandate remained limited regarding North and South Kivu.[106] As a result, the Security Council adopted resolution 1565 (October 2004), providing MONUC with a more robust mandate and increased troop size to 16,700. Approximately 5,500 troops were redeployed to the Kivus and Ituri and the MONUC Eastern Division was established.[107]

With an impending 2006 elections, the Security Council by its resolution 1592, March 2005, authorized MONUC "to use all necessary means within its capabilities (. . .) to deter any attempt at the use of force to threaten the political process and to ensure the protection of civilians under imminent threat of physical violence, from any armed group."[108] MONUC adopted a proactive strategy, shifting from reaction to pursuit in 2005, resulting in aggressive action against possible spoilers. After successful democratic elections in 2006, violence re-erupted in the eastern DRC and security conditions deteriorated, particularly in the volatile provinces of Kivu and Ituri. MONUC's proactive attitude heralded the beginning of widespread reprisal killings of civilians in the area. However, following the 2006 elections, the mission's protective activity became more limited with the new government assuming responsibility of civilian protection. This was articulated in the UN Secretary-General's 23rd report on March 2007, where he declared that the mission's role from then on would be mainly restricted to supporting the FARDC. MONUC was vastly overstretched then and the Spanish General Vicente Diaz De Villegas, who succeeded General Babaçar Gaye as the new MONUC Force Commander, resigned as a consequence of what he called a "dangerous mission creep."[109] MONUC faced challenges in protecting civilians as it failed to prevent the killing of civilians in North Kivu's capital Goma. Due to the prevailing political environment and increasing government criticism, the Security Council, by its resolution 1925 on July 1, 2010, renamed MONUC the United Nations Organization Stabilization Mission in the Democratic Republic of the Congo (MONUSCO) to reflect the new phase reached in the country. MONUSCO was authorized to use all necessary means to carry out its mandate relating to, among others, the protection of civilians, humanitarian personnel, and human rights defenders under imminent threat of physical violence and to support the government of the DRC in its stabilization and peace consolidation efforts. Nevertheless, civilian protection remains the mission's first priority. MONUSCO has also failed to protect civilians from attacks by spoilers. For example, MONUSCO peacekeepers did not intervene during the Kibua–Mpofi mass rapes by Mai Mai Cheka rebels despite located only 30 kilometers from the attacked area. MONUSCO faces resource constraints and operational limitations that include lack of telephone coverage obstructing communication with the population in the area amidst a climate of distrust between the population and the peacekeepers.[110] In May 2011, the international cooperation minister of DRC expressed his government's desire of an "orderly, progressive withdrawal" of MONUSCO due to "normalization" of DRC's relations with neighboring countries and containment of rebels to a "few isolated zones."[111]

In June 2011, India decided to withdraw all of its four Mi-35 combat helicopters from MONUSCO and eventually cease peacekeeping operations in DRC.[112] Following this, the United Nations declared the intention to withdraw about 2,000 peacekeepers.[113] Meanwhile, MONUSCO on November 15, 2012, supported government forces with its helicopter gunships in repelling attacks against civilians in eastern DRC by 64 (March Movement 23 (M23)) fighters.[114] MONUSCO continues to focus on protecting civilians in the east amidst mixed results.

Analyses

MONUC/MONUSCO is designed in robust terms. It is equipped with heavy weapons such as attack helicopters and also authorized to "use all necessary means . . . to deter any foreign or Congolese armed groups from attempting to use force to threaten the political process and to ensure the protection of civilians under imminent threat of physical violence."[115] MONUC/MONUSCO also benefits from major power participation, especially from European Union forces. For example, in September 2003, the European Union force (EUFOR) led by Germany and France conducted in Ituri "Operation Artemis" to stabilize the east capital of Bunia in September 2003. Again, from July to December 2006, EUFOR helped MONUC to maintain stability, especially during the elections.[116] Operation Artemis, deployed under Chapter VII to enforce peace, successfully rescued the peace process by restoring peace there because it was composed of well-trained soldiers and had access to adequate resources. With large force size, robust rules of engagement, possession of heavy weapons and major power participation, MONUC has the potentiality of constituting a formidable barrier to intentional killing of civilians by the spoilers. MONUC, in view of these characteristics, possesses enough greater firepower, stronger offensive capabilities enough to constitute a formidable opposition to alter the calculations of the rebels by raising the costs of continued fighting and benefits of opting for peace. Buoyed by its stronger firepower and offensive capabilities, MONUC/MONUSCO and Congolese army (FARDC) supported by attack helicopters and Special Forces regularly undertake joint military operations to flush out foreign groups such as the Rwandan FDLR, the Allied democratic Forces, and National Army for the Liberation of Uganda (ADF/NALU) in east DRC that have been killing civilians and raping women.[117]

MONUC's mobile deployments, large-scale protection operations, and tailored military patrols have had a clear impact and have contributed substantially to the containment of military violence, the prevention of massacres, and the safe delivery of humanitarian assistance in a complex and often chaotic environment. MONUC/MONUSCO's tough approach has been reasonably successful in strengthening security in some parts of the east and safeguarding the elections in 2006. Also, since the initiation of the offensive actions coupled with disarmament, demobilization, repatriation, reintegration, and resettlement (DDRRR) programs for all foreign-armed groups, some 13,000 foreign fighters and families have returned voluntarily to Rwanda, Burundi, and Uganda, reducing the harassment and killing of the civilians.[118] The death rate has decreased in the other parts of DRC except east DRC, and civilian deaths are more attributable to hunger and diseases than intentional civilian killings.[119] A more active, aggressive form of peacekeeping seems to be achieving results in some parts of DRC partly driven by the very size of MONUC's force, heavy equipment, robust rules of engagement, and major power participation that constitute a formidable barrier to the spoilers' capacity to kill civilians as a mechanism of pursuing their political agenda.[120]

In spite of a measure of success achieved, MONUC/MONUSCO has far from fully implemented its mandate of protecting civilians.[121]

MONUC/MONUSCO's offensive, operational capabilities and strong firepower should have been able to raise the costs of continued fighting and benefits of peace for the spoilers high enough but MONUC/MONUSCO's efforts to constitute a formidable barrier has been obstructed by a number factors.

A major challenge to the mission's ability to protect civilians is the mismatch between the available peacekeeping forces and the size of the opposition and their operating environments. Although MONUC/MONUSCO's 17,000 strong force is popularly touted as the largest UN peacekeeping force but it is small relative to the size of the country and the strength of the oppositions. MONUC constitutes the same size force as UNAMSIL had in Sierra Leone, a country that is 1/24th the size of the DRC. It is also estimated that the FDLR rebel faction, mostly Rwandan-based Hutu *genociders* based in eastern Congo, has a force strength of 15,000 armed rebels. As a result, MONUC/MONUSCO is not fully able to carry out its task of protecting civilians due to an insufficient number of troops comparable to the size of opposition. The profusion of at least 15,000 strong hostile armed factions in the Kivu and Ituri Provinces in east DRC makes it extremely difficult for MONUC to effectively fulfill its protection mandate since MONUC's protection mandate covers all of DRC and therefore it has to spread thin on the ground. MONUC therefore lacks the appropriate comparative force size and offensive capacity to deter the rebels from attacking the civilians. MONUC faces human resources constraints[122] because few developed countries are willing to deploy their soldiers to the DRC. Most of the forces are from developing countries, which sometimes have poor human rights records that undermine the legitimacy of the mission as an honest peace broker. Moreover, as each Member State is responsible for training and equipping its troops, most of the troops from the developing world fall short of the requisite training, preparation, motivation, and equipment for robust missions compared to Western troops. The bulk of MONUC's 17,000 peacekeepers are from India (4,300 deployed in North Kivu), 3,500 from Pakistan (South Kivu), and 1,300 from Bangladesh. As a result of the this constraint, when thousands of rebels led by the dissident Officer Laurent Nkunda and Colonel Jules Mutebusi attacked Bukavu in 2003, MONUC had only 800 peacekeepers and therefore failed to withstand them. As a matter fact, many peacekeepers fled into their compounds, leaving the population to defend themselves.[123] HRW cites, among others, logistical and technical constraints as the most significant causes of MONUC's failure to protect civilians in Kiwanja.[124] The mission would need about 100,000 peacekeeping personnel to be able to effectively provide security.[125] The mission also faces severe material resource constraints such as helicopters and interpreters, which are vital in protecting civilians. Due to the language and cultural barriers between peacekeepers and civilians, there is an information gap and peacekeepers face difficulty in obtaining appropriate intelligence to proactively protect civilians under attack.[126] There is therefore a need for interpreters if the mission is to successfully execute its mandate of civilian protection.

In spite of the heavy support burden shouldered by MONUC, the mission's primary responsibility continues to be the protection of civilians, a task not easy to accomplish in the vast, hilly, densely forested eastern region, with just one MONUC soldier for every 123 square kilometers. The difficult terrain, broken infrastructure, and the vast size of the country have affected MONUC's monitoring capacity and ability to respond swiftly to attacks on civilians. The country has only 300 miles of roads despite being the size of Western Europe. The few roads and difficult terrain hinder the mobility of armored personnel carriers and other heavy military equipment keenly required to deter the spoilers. Operations in such an area of natural barriers such as mountains and swamps affect peacekeeping forces' ability to monitor human rights and achieve operational success, especially as spoilers might have an advantage of knowledge of and maneuverability in such terrain. Nonetheless, huge protection gaps continue to exist. Violence, theft, and intimidation persist in the many areas that MONUC soldiers simply cannot cover. MONUC is absent in remote inaccessible villages and mining areas such as Bisie that are the most problematic areas with bulk of the attacks taking place.

Furthermore, the porous Congolese borders that allow easy infiltration by interested external actors such as Rwanda and Uganda with vested political and economic interests compound MONUC's problems.[127] The governments of Uganda and Rwanda accuse the UN and Congolese troops of failing to control the rebel groups that occasionally launch attacks across their borders. As a result, both countries continue to intervene covertly in DRC's internal affairs, launching attacks against these rebels as well as civilians.[128] Rebel groups such as Nkunda's group receive cross-border re-supply of arms and personnel from Rwanda. DRC still harbors large armies of foreign Ugandan and Rwandan rebel factions in the east of DRC that prey on civilians. Between 8,000 and 12,000 FDLR, Rwandan Hutu rebels, and LRA, Ugandan rebels, still roam the hills of South Kivu, offering both Rwanda and Uganda a real incentive and an ideal excuse to remain deeply involved in the DRC conflict, launching attacks against these rebel groups as well as civilians. The severity and duration of the conflict is heightened by foreign conflict parties' strategic economic interests in Congo's resources and support for various rebel groups. A UN panel of experts' investigation of illegal exploitation of diamonds, cobalt coltan, gold, and other lucrative resources in the Congo in April 2001 accused

Rwanda, Uganda, and Zimbabwe of systematically exploiting Congolese resources and recommended the Security Council impose sanctions.[129] The accusations were rejected by both Rwanda and Uganda, denying the involvement of their senior political and military figures in illicit trafficking of plundered resources.

The support of the numerous rebels in east DRC by their home governments, such as Rwandan governments' sponsorship of support for Nkunda's group, has emboldened them and increased their risk tolerance level and their propensity to attack both civilians and FARDC/MONUC forces. The rebels' formidable resistance and their tenacity to risk, coupled with other mitigating factors such as limited mission strength, have lowered the casualty tolerance level of some risk-adverse MONUC contingents, especially those with limited strategic interests in the conflict. The risk-averse contingents, informed by the calculation about risk to troops, strive to minimize the exposure of their troops to danger and avoid deployment in dangerous parts of the mission theater. The Kisangani massacre of more than 180 people near a UN camp in May 2003 and the massacre of civilians in Kiwanja[130] and Bukavu by the Laurent Nkunda-led rebels where many peacekeepers fled into their compounds, leaving the population to defend themselves,[131] illustrate the fear of some risk-averse MONUC contingents to expose their troops to danger and the failure to protect civilians.

A more serious dilemma facing MONUC forces is the absence of conceptual clarity and articulation on clear guidelines on the use of force by mission forces in this "gray area" between "peacekeeping" and "enforcement." This hardly presents coherent guidelines to military commanders and field practitioners on the use of force culminating in differential interpretations and levels of force used by different troop contingents within the same mission. With the absence of conceptual clarity on use of force, we witness differential interpretations and possible conflict between United Nations' rules of engagement and that of contributing nations on the use of force. As a result, participating contingents, even in robust forces, are affected by the "phone home" syndrome in which participating contingents seek domestic legal guidance in respect of the use of force. Contingents' reliance on their domestic legal systems to guide the use of force emanates from the participatory precondition that contributing states have disciplinary jurisdiction over its troops and not the United Nations. Clashes between the missions' national contingents' approaches, doctrines, and interpretations on the use of force within a mission can lead to incoherent and ineffective peace operation. The mission's mandate to use "all necessary means" including force to protect civilians then loses its resonance when UN troops are constrained by their national laws.[132]

The differential applications of force within the same mission also raise questions on the missions' legitimacy. This is the problem MONUC faces in DRC. MONUC peacekeepers have been observed to adopt a "hesitative attitude towards the use of force and narrowly interpret the rules of engagement to use force only for self-defense or force protection."[133] There seems to be "a sharp decline in firing engagements by contingents since 2006, indicating an apparent growing aversion to the use of force, despite continuing threats to civilians."[134] According to MONUC's lessons learned, during the 2003 conflict in Bunia, the Uruguayan MONUC battalion (URABATT) deployed there to protect civilians failed to do so because they thought they were to guard UN infrastructure and personnel and more than 400 civilians were massacred in two weeks.[135] Former Force Commander (FC) Jan Isberg, in an internal MONUC report, explained that URABATT failed to protect civilians because they thought they were not operating under Chapter VII and therefore not allowed to use force except in self-defense. Rather than protecting the civilians, URABATT command sought authorization from the Uruguayan parliament.[136]

In some cases MONUC forces faced the dilemma as to what is the minimum force that should be enough to deter spoilers while at the same time avoid tipping over into the sphere of enforcement. During the Bukavu crisis of May–June 2004, MONUC forces had access to robust resources such as attack helicopters but did not use them. The crisis illustrates the differential and confused interpretation on the use of force between the mission's military and political components. While MONUC's military command wanted to go on the offensive, DPKO preferred to stay out of Congo's "internal affairs."[137] The massacre of more than 180 people in Kisangani in May 2002 near a UN camp also illustrates United Nations' tendency to compromise on civilian protection fearing that the use of force might offend parties to the peace process.[138] With an impending 2006 presidential elections in DRC, the international community resolved to safeguard the elections and the UNSC by resolution 1592, March 2005, authorized MONUC "to use all necessary means within its capabilities (. . .) to deter any attempt at the use of force to threaten the political process and to ensure the protection of civilians under imminent threat of physical violence, from any armed group." MONUC accordingly adopted a proactive posture and aggressively deployed offensive operations against possible spoilers in the East to safeguard the elections.[139] MONUC's proactive use of force, however, led to reprisals against civilians and increased the risk of collateral damages. The DPKO, contrary to MONUC's protection mandate in SC resolution 1592, March 2005, wished for all offensive actions to stop. The DPKO acknowledged the mission's protection mandate but also claimed that "it was not the peacekeepers' role to go on the offensive and take out the militias preying on civilians. We are not engaged in war."[140] The danger facing MONUC is that too little could lead to mission failure to protect civilians while too much could be deemed excessive and elicit a violent backlash from spoilers and global condemnation. In east DRC, MONUC's failure on several occasions to use force to protect the civilian population from indiscriminate attacks led to violent protests and attacks against MONUC forces. Such provocative attacks on UN forces undermine the missions' reputation for robustness—the UN forces will not fire back against civilians, which gives the false impression that their firepower is essentially symbolic. The cases of DRC illustrate this dilemma. There are indications that MONUC over time has experienced a loss of legitimacy among the civilian population in eastern DRC, partly because of its failed "robust" stance.[141] The protests prevented MONUC to carry out its duties and consequently led to a reduction in the number of patrols in the region of Kiwanja.[142]

MONUC is also limited by severe constraints inherent in the UN system, particularly the characteristic heavy UN bureaucratic and hierarchical malaise of top–down structure with complex procedures where every action requires approval from the top. This sometimes demobilizes the operational capacity of the field troops due to delays on operational decisions by DPKO. MONUC forces are therefore compelled to adopt a reactive posture rather than a proactive attitude. They do not engage in preventive operations but repair the damage. Some risk-averse contingents also capitalize on United Nations' structural hierarchical and bureaucratic intricacies as an excuse for inaction.[143] The military structure of UN peacekeeping missions also constraints MONUC's operational capabilities in protecting civilians. UN peace missions lack an integrated chain of command and common procedures. Even though different national contingents are deployed under one Force Commander, the participatory preconditions allow each contingent to report to its contingent commander who is first accountable to his own government, before obeying the Force Commander. Another problem is that contributing states have disciplinary jurisdiction over their troops and not the United Nations. The emergent constraint is that national interests supersede mission interests and strongly influence the contingents' performance during military operations.[144]

MONUC's joint operations with and logistic support for the Congolese Armed forces (FARDC) against the other conflict parties have also undermined its impartiality and legitimacy as an honest broker. FARDC is discredited for massive human rights violations and MONUC is accused of condoning the condemnable FARDC's immoral and illegitimate acts. The mission is considered by some civilians as FARDC's accomplice.[145] Consequentially, the affected conflict parties, especially the Hutu-dominated FDLR, have refused to cooperate with the United Nations and increased their attacks on civilians. The renegade general Nkunda, leader of the Tutsi rebel faction, who claims to be protecting the minority Congolese Tutsi population from FDLR, accused the MONUC of siding with the Congolese government forces and warned of retaliation.[146] In this regard, a failure to distinguish between impartiality and neutrality is seen as one of the key errors made by MONUC by undertaking joint military operations with the Congolese Armed Forces against some factions in the Congolese conflict.

MONUC's involvement in sex, bribery, corruption, and illicit trading scandals has undermined its credibility and reputation as a legitimate peace broker thereby eliciting non-cooperation from some conflict parties. MONUC's military and civilian staff members were reported in 2004 to have committed serious acts of sexual exploitation and abuse resulting in a public scandal.[147] The BBC in 2005 reported that Pakistani MONUC peacekeepers in Mongbwalu swapped arms for gold with Nationalist and Integrationist Front (FNI) militia leaders. International newspapers reported in 2007 that Indian peacekeepers in Goma had swapped arms for minerals with rebel movement CND.[148] Africa Confidential also reported in May 2008 that some personnel of the Indian contingent swapped ammunition for ivory with rebels, fraternizing with the FDLR and failing to support its disarmament, buying marijuana from the FDLR, and exchanging UN rations for dollars and gold with the rebels.[149] Reports indicate that the UN peacekeepers are being investigated over allegations of trading weapons and military intelligence with the FDLR rebels in return for gold.[150]

In conclusion, MONUC by its robust design with adequate firepower and offensive capability possesses the capacity of constituting a formidable barrier to the killing of civilians by the conflict parties in the civil conflict in DRC. Despite a measure of success, it has far from fully implemented its mandate of protecting civilians especially in east DRC. There are reports of civilian killings and other human rights violations. MONUC's efforts to protect civilians have been undermined by the presence of Rwandan and Ugandan rebels in the east of DRC, which predictably engenders the involvement of proxy Rwandan and Ugandan troops in the conflict. Compounding the problem is DRC's porous borders that facilitate external actors' unchecked intrusion and attacks on civilians as well as providing support for some conflict parties. Worse still, the large size of the opposition, the vastness of the country, and the poor topography and infrastructure constrain MONUC's monitoring and patrolling abilities. MONUC's problem is worsened by its partiality and scandals that have elicited noncooperation from certain conflict parties. MONUC is severely limited by material resource constraints, the characteristic heavy UN hierarchy and bureaucracy malaise, and the lack of conceptual clarity and agreements on the use of force. Though a robust mission, MONUC has produced mixed results of failures and success due to both structural and operational constraints confronting it.

Conclusion

This chapter has examined the cases of traditional and robust peacekeeping to determine which works better in reducing deliberate civilian killings. These cases were examined in the context of the research questions, hypotheses, and the generated *formidable barrier* model of peacekeeping success, which posits that for a UN mission to stop conflict parties from killing civilians, it must be able to raise the costs of continued fighting and benefits of choosing peaceful negotiations of conflict. In other words, the mission should have the capacity to alter the calculations of conflict parties to opt for peace and avoid attacking civilians. I observe that UNAMSIL and MONUC, key representative cases of robust peacekeeping in Africa in comparison to UNAVEMs and UNAMIR, key representative cases of traditional peacekeeping in Africa, constituted formidable barriers to the killing of civilians by the conflict parties. UNAMSIL and MONUC are characterized by large force size, use of heavy weapons, possession of rules of engagement that permitted the use of force to protect civilians and major power participation, and an enhanced firepower and offensive capacity enriching their capacity to protect civilians from attacks by the spoilers. Despite MONUC's success, its capacity was undermined by certain factors such as opposition size, vastness and poor topography of the country, and partiality and malpractices of the mission.

Contrary to the robust cases, the traditional cases were characterized by limited force size, restricted rules of engagement that limited the use of force only in self-defense, use of small arms, and no major power participation, which did not constitute formidable barriers to prevent the killing of civilians by the spoilers. By their designation, the traditional cases lacked firepower and offensive capability needed to alter the calculations of conflict parties by raising the costs of continued fighting and benefits for the conflict parties.

The study therefore supports the theoretical argument, presented in Chapter 3, of the likely effect of large force, heavy arms and robust rules of engagement, and major power participation on lowering intentional civilian killings. The absence of these factors in our traditional peacekeeping cases seems to have adversely affected the missions' capacity to stave off intentional civilian killings. The case study also captured certain potential factors, excluded in my quantitative analyses, which impacted on intentional civilian killings. The factors: late deployment; mission force partiality and crimes; porous border; poor topography and country size; risk tolerance level of both mission forces and spoilers; UN bureaucratic and hierarchical structures and military command system; lack of conceptual clarity and agreements on the use of force; and external parties strategic interests in the conflict have effects on the operational capacity of some missions in reducing intentional civilian killings.

7

The Politics of Civilian Protection—The Case of Darfur

The international community of states promised "never again" would they be bystanders when moral atrocities are committed against innocent civilians following the United Nations' failure to intervene robustly in Rwanda to prevent or halt the genocide. British prime minister Tony Blair promised that "if Rwanda happens again we would not walk away as the outside has done many times before," and insisted that international society had a "moral duty" to protect civilians from atrocities.[1] The United Nations' abysmal response to the Rwandese genocide weakened the faith the global community had in its commitment to protect civilians in violent civil wars. The organization thus acquired the stigma of moral failure and cowardice in the face of evil. Secretary-General Kofi Annan described the United Nations' inaction as a "sin of omission."[2] Fear of its operational paralysis and concomitant declining legitimacy and reputation gained primacy in subsequent UN debates. Discussions fixated on the United Nations' efficacy to fulfill its obligation of civilian protection against moral atrocities. This ignited the reform processes culminating in the development of international protective norms such as the "responsibility to protect" that transforms sovereignty as a right to a responsibility. Responsibility to protect obliges states to protect their citizens from avoidable catastrophe, but failure to act triggers international obligation to ensure that the dignity of the human person is respected using all necessary means, including force. These reformative developments marked the rebirth of heightened optimism and expectations of the United Nations' fulfillment of its civilian protection obligations. Darfur massacres tested the United Nations' commitment to civilian protection against egregious human rights violations. Unfortunately, the world failed to go beyond "never again" rhetoric to protect Darfurians despite overwhelming evidence of atrocities that met the thresholds that trigger robust intervention. Governments that overwhelmingly endorsed the international norm of "responsibility to protect" willfully ignored their moral duty to protect Darfurians caught in the throes of violations by their own government and willful allies, the Janjaweed. The United States called the Darfur killings genocide and the United Nations and Western nations described it as the "most serious humanitarian emergency."[3] Other NGOs, such as Physicians for Human Rights and Justice Africa, called the crisis genocide.[4] The UN Security Council, the EU, and a variety of NGOs (Amnesty International, Human Rights Watch)—all acknowledged the complicity of the Sudanese government in large-scale crimes against humanity and ethnic cleansing in Darfur.[5] The Sudanese president El Bashar and his cronies have subsequently been indicted by the International Criminal Court for war crimes and crimes against humanity that are criteria for humanitarian intervention. UN concern about the situation in Darfur was strongly expressed by Secretary-General Kofi Annan when he cautioned the United Nations against inaction in a speech before the UN Human Rights Commission on the eve of the tenth anniversary of the Rwandan genocide. Annan said that events in Darfur left him "with a deep sense of foreboding. Whatever terms it uses to describe the situation, the international community cannot stand idly by . . . [but] must be prepared to take swift and appropriate action. By action in such situations I mean a continuum of steps, which may include military action."[6] Despite the overwhelming evidence of crimes against humanity, complicity of the Sudanese government, and global demand for robust intervention, the international community failed to contemplate this action. Darfurians were cheekily handed a hybrid mission that lacked the political will and capacity to stop the massacres. One thorny question is why the world denied Darfurians a robust response as happened in Iraq and Kosovo, where the severity of killings paled in comparison to those in Darfur.

Darfur exposed serious shortfalls in the United Nations' ability to use force and political will to halt ethnic cleansing and protect civilians. It demonstrated that UN ambitions to robustly protect civilians in violent conflicts is hampered by the neorealist national interests of the five major powers of the Security Council that determine protective interventionist decisions. The UN response to Darfur killings reveals that decisions for humanitarian interventions to protect civilians are grounded in national interests rather than genuine altruism. Interventions to protect civilians are purely and politically motivated by the neorealist interests of the P-5. Politicization of humanitarianism in the Security Council has hijacked the world's moral duty of civilian protection. The pallid response by the UN Security Council and stakeholders, including the African states, has raised doubts about the commitment of the global community to honor its responsibility to protect civilians from catastrophic sufferings in violent wars. The collective yawn in the face of the Darfur massacres has blighted the world's commitment to its "never again" to dither when heinous crimes are committed against defenseless civilians. This chapter explores why the strongest advocates of "responsibility to protect" failed to protect civilians suffering at the hands of their own government in Darfur. I argue that politics within the UN Security Council and the concomitant divisiveness fuelled by neorealist strategic national interests have led to the humanitarian tragedy in Darfur. The chapter proceeds in three parts. It begins by providing an overview of the legal and moral justifications for robust interventions to protect civilians. This is followed by an overview of the Darfur crisis and a critical analysis of international responses to factors that dissuaded the United Nations, the West, and Africa from intervening robustly to protect Darfurians.

Legal and Moral obligations to Protect

Robust interventions by the international community to save defenseless civilians from heinous crimes can be justified by two complementing logics of "moral" and "legal" obligations. International laws protecting human rights oblige Member States to protect civilians from atrocities against them. The United Nations proclaim in Article 1 of its Charter the protection and promotion of human rights as a purpose, and Articles 55 and 56 of the Charter impose obligations on members to protect and promote human rights. In Articles 55 and 56 of the Charter, all members "pledge themselves to take joint and separate action in cooperation with the Organization" to achieve the purpose in Article 1 of the Charter. States' memberships reflect their subscription to the principles of the UN Charter and commitment to promote and protect human rights and robust intervention to protect human rights is consistent with the purposes of the United Nations. Both the General Assembly and the Economic and Social Council (ECOSOC) are charged to make recommendations to promote human rights in Articles 13 and 62 of the Charter. Of special significance is the Universal Declaration of Human Rights that proclaims the universality of human rights permitting collective responsibility to curb atrocities. Article 39 empowers the Security Council "to determine the existence of threats to the peace and breach of the peace." In the same article, the Council declares armed intrastate conflict a threat to international peace and security. Article 24(1) of the UN Charter confers responsibility on the Security Council to maintain international peace and security using either Chapter VI, peaceful means, or Chapter VII, robust measures including the use of force if Chapter VI measures prove inadequate.

Customary international laws impose legal obligations on all states to protect fundamental values of the international community. The International Court of Justice (ICJ) describes the customary international law in Article 38(1)(b) as "evidence of a general practice accepted as law."[7] These are international legal obligations arising from a general and consistent established state practice that states cannot derogate from them through international agreement.[8] In recent years states have developed with attendant *opinion juris* (legal conviction) on a set of fundamental values worthy of protection that indisputably are peremptory norms or *jus cogens*.[9] Article 53 of The Vienna Convention on the Law of Treaties of 1969 (and 1986) describes a peremptory norm, or *jus cogens*, as a "norm accepted and recognized by the international community of States as a whole as a norm from which no derogation is permitted."[10] Respect for *jus cogens* preserves the fundamental values of the international community and offers protection against atrocious crimes such as genocide, war crimes, and crimes against humanity. Most delegates at the Vienna Diplomatic Conference on the Law of Treaties stated that the fundamental principles of international humanitarian law such as genocide belong to *jus cogens*.[11] The ICTY in *Zoran Kupreskic* et al. in 2000 referred to "most norms of international humanitarian law, particularly those prohibiting war crimes and crimes against humanity as peremptory norms."[12] Peremptory norms such as genocide impose community obligations *that* are *erga omnes*—obligations that apply to all states [13] as well as universal jurisdiction. *Erga omnes* norms include prohibitions against piracy, genocide, slavery, and torture.[14] *Erga omnes* obligations protect universal critical rights and prevent their breaches. Article 2(c) of the Genocide Convention describes genocide as "inflicting the conditions of life calculated to bring about the physical destruction of a group."[15] Genocide is a crime under international law; states undertake to prevent it and to punish the guilty.[16]

The principle of universal jurisdiction provides a legal basis under which robust intervention to halt atrocities against defenseless civilians, as in the cases of Rwanda and Darfur, is morally permissible.[17] Universal jurisdiction is the power of states to claim jurisdiction over an accused, irrespective of where the alleged crime was committed and regardless of the accused's nationality or country of residence. Secretary-General Kofi Annan clearly articulated this obligation when he stated, "Crime of Genocide puts all States under an obligation to 'prevent and punish' this most heinous of crimes." It also allows them to "call upon the competent organs of the United Nations" to take action for this purpose.[18] Crimes prosecuted under universal jurisdiction are *jus cogens,* such as genocide, war crimes, and crimes against humanity. The principle of universal jurisdiction is structurally linked to the universal conception that some international norms are *erga omnes,* or owed to the entire world community, and that certain international law obligations are binding on all states.[19] Article 4 of the African Union's Constitutive Act, for example, establishes the right of the African Union to intervene against Member States in the case of grave circumstances, namely: war crimes, genocide, crimes against humanity, and serious threats to legitimate order in order to restore peace and security.[20]

There is also an *opinion juris* in promising state practice today to deploy robust peacekeeping missions, such as UNAMSIL in Sierra Leone, mandated to protect civilians from human rights violations in violent civil wars. Robust mandates often denote the United Nations' legal basis and intention to use force to defend its personnel, humanitarian actors, and civilians under threat of physical harm by conflict spoilers. Annan signaled a shift in rhetoric by declaring the protection of civilians a humanitarian imperative.[21] Consequently, two UNSC thematic resolutions—S/RES 1265 (1999) and S/RES 1296 (2000)—indicated that the targeting of civilians constituted a threat to international peace and security, imposing obligation on the Council to take necessary remedies under Chapter VII of the Charter. When the Council established a robust UNAMSIL mission through S/RES 1270, the Council made direct reference to its new legal powers by mandating the mission to take necessary action "to afford protection to civilians under imminent threat of physical violence."[22] Certain conceptual developments contributed to the shift in the UN peacekeeping operational paradigm from traditional to robust. The 1992 Agenda for Peace and its 1995 Supplement challenged notions of "absolute and exclusive sovereignty," pressed for greater human security and affirmed the United Nations' right to robust Chapter VII intervention in a state to protect citizens' human rights. In *Two Concepts of Sovereignty,* Annan echoed "the need for timely intervention by the international community" to halt atrocities in a state. The 2000 Brahimi Report also recommended more assertive and deterrent mission force to halt human rights violations. Since the first landmark UNAMSIL resolution 1270 (1999), most UN missions have been authorized under Chapter VII and have been provided with robust rules of engagement (ROE) to "take the necessary action to ensure the security and freedom of movement of its personnel and to afford protection to civilians under imminent threat of violence . . . "[23] The United Nations' historic operational paradigm shift is also reflected in *The Report of the High-level Panel on Threats, Challenges and Change, A More Secure*

World: Our Shared Responsibility of 2004, which highlighted the role of the United Nations in facing the security challenges of the future. The report, informed by events in Iraq, Afghanistan, and Darfur, endorsed the acceptance of R2P as an emerging norm and recommended criteria for international military intervention.

Humanitarian provisions of the laws of armed conflict, international human rights laws, and norms such as responsibility to protect constitute the fundamental basis of moral obligations to intervene in violent civil wars to protect civilians from heinous crimes against them. The principles of human rights and concern for human dignity and the rule of law morally oblige the United Nations to intervene robustly rather than stand by when the most egregious violations, shocking the conscience of humankind, are committed against defenseless people who the UN Charter proclaims to protect. The norms of extra-judicial execution, torture, and genocide clearly stated in major human rights instruments provide both legal and moral obligations to punish perpetuators.[24] Intervention to stop a barbarous cruelty as in the Darfur case is morally right and of higher pedigree than the concept of impartiality. Where there is blatant disregard for human dignity by conflict parties, continued impartiality by the United Nations can, in the best case, result in ineffectiveness, and in the worst may amount to complicity with evil.[25] The universal acceptability of moral obligation to stop heinous crimes is reflected in the international community's, particularly the United States', admittance of guilt following its refusal to honor its moral responsibility to stop the Rwandese genocide. President Clinton's atonement speech to Rwandese survivors during the tenth anniversary of the genocide is indicative of the guilt.

In its simplest form, robust intervention by UN forces to protect civilians demonstrates that society does not brook behavior that breaches respect for human dignity, and signals to the victims that their plights will not go unnoticed. It signals to the citizens of the global community that everyone is important and not disposable, and that offenses against their inherent human dignity will not be tolerated. It therefore concretizes the rule that interventions using force in the interest of humanity are admissible. Robust intervention shows the importance society assigns to the norms that prohibit genocide, war crimes, and crimes against humanity. If impunity pervades without response from the international community, what guarantees that it will not be applied in the future? Impunity unchecked at infancy may lead to a cycle of impunity into the future, undermining effective resolution of underlying differences, reconciliation, and peacemaking. Robust opposition to spoilers instills voluntary respect for just conduct by discrediting inhumane or unjust conduct. This ultimately encourages habitual conformity with the law.

The United Nations basically upholds the prosecution and punishment for war crimes and crimes against humanity as well as the protection of human rights. In relation to these, a robust intervention by a deterrent force with the capacity to protect human rights in civil conflict theatres increasingly appears to be indispensable in upholding the rule of law and human rights on a global scale, especially when elementary forms of humanitarian and human rights are blatantly trampled upon. Furthermore, society needs to respond to violations to restore human dignity and respect for rule of law. Failure to respond to egregious crimes such as genocide and extra-judicial killings is a moral failure that sets precedence with serious undermining consequences for confidence in the rule of law and personal safety in conflict zones.

The moral authority of a robust UN interventionist force to save civilians is derived from the international community's commitment to human rights protection and abhorrence of standing idly by while human rights violations take place, as happened in the Rwandan genocide. Following the failures in Rwanda and Bosnia, and the subsequent apology by Clinton, the international community seems to be advocating an activist stance against genocide, war crimes, and crimes against humanity. Since then, the notion of a collective responsibility toward individual's human rights has seeped into the culture and identity of the international community. This humanitarian posture is reflected in the establishment and overwhelming support for the international norm of "Responsibility to Protect" (R2P) in 2005 by UN Member States. The norm transformed state sovereignty from a right to a responsibility.[26] R2P focuses on preventing and halting genocide, war crimes, crimes against humanity, ethnic cleansing, and mass atrocity crimes.[27] The core tenet of R2P is that state sovereignty entails responsibility to protect their citizens from these heinous crimes but when unwilling or unable to protect them, the moral obligation of the international community is triggered using all necessary means including force to intervene and protect populations through the collective action of the United Nations. With this, there appears a resurgence of global interest in ensuring respect for the moral dignity of the human person. With such a global consensus, there is a moral obligation to act when these values are violated. The *United Nations Reforms 2005 Report* acknowledges that the major accomplishment for human rights was the normative progress in the use of force in the face of conscience-shocking events such as crimes against humanity and war crimes. The international norm of R2P was reaffirmed in SC resolution 1674 (2006) on the protection of civilians in armed conflict. The African Union officially adopted R2P in March 2005 and committed to strengthening its implementation in a resolution of the African Commission on Human Rights and People's Rights in November

2007. Formal codification of the R2P continued within the context of the Darfur crisis with the passage of SC resolution 1706 (2006) in August 2006. Humanitarian provisions of the international norm of "responsibility to protect," overwhelmingly supported by UN Member States, constitutes a moral obligation to protect civilians in violent civil wars. However, the articulation of the R2P in the Security Council's resolutions did not translate into robust intervention to stop human suffering in Darfur. Darfur was handed a weak hybrid model of peacekeeping mission that lacked the capacity to protect civilians and itself. Although resolution 1556 was passed under Chapter VII of the UN Charter, the Security Council chose not to assume responsibility for alleviating human suffering in Darfur by not authorizing a humanitarian intervention. The question is: Why were Darfurians denied protection by the international community despite the overwhelming evidences of atrocities that have met the criteria of robust intervention? Answering the question illuminates considerations other than legal or moral that informs decisions to intervene robustly in violent civil wars to protect civilians under siege of their own government as in the case of Darfur.

Darfur Conflict

The conflict is rooted in the structural inequality between the developed center of the local Arabic-speaking pastoral class supported by the Sudanese government and the relatively underdeveloped peripheral areas of diverse indigenous inhabitants of the Darfur region that resulted in historical animosities shaped along ethnic fault lines. The current Darfur conflict is between the Sudanese military and the Janjaweed, a militia group from the camel herding nomad tribes of the northern Rizeigat on one side and a variety of rebel groups from the land-tilling Fur, Zagaghawa, and Massaleit ethnic groups. Almost all the combatants and victims in Darfur are Muslims, but there is an Arab–African dichotomy. The Sudanese military and the Janjaweed have Arab origins while the rebel groups for Darfur are black Africans.[28] The Darfur war started in February 2003 when two rebel groups, the Sudan Liberation Army (SLA) and the Justice and Equality Movement (JEM) forces, attacked government military installations based on complaints of decades of recurrent government exploitation, political marginalization, economic neglect, and regional lack of development. The government responded by deploying its forces (notably the air force) in joint counterinsurgency operations with the *Janjaweed* militias (a popular umbrella term encompassing outlaws/ bandits, Arab tribal militias, and the Popular Defence Forces). The Khartoum government-backed Arab militias (*Janjaweed*) attacked non-Arab villages; engaged in burning homes, killings, civilian mutilations, abductions, forced expulsions, systematic sexual violence; and deliberate destruction of crops, livestock and important cultural and religious sites. Though the Khartoum regime denies control over these militias, the *Janjaweed* are certainly not denying their pro-government stance. And refugees who have fled into neighboring Chad report that Janjaweed raids often follow government aerial bombardments with helicopters on villages. The war has resulted in a spiral of violence and collective savagery reminiscent of the dark ages with over 500,000 dead and 2.5 million people displaced.[29] Some crossed into Chad, while the majority remained internally displaced within camps in Darfur, suffering from incessant Janjaweed attacks and exploitation, and sealed off from most international relief agencies.[30] An agreement signed in November 2004 under the auspices of the African Union paved the way for the deployment of the ill-equipped and poorly funded African Union forces in Darfur. These forces lacked the capacity and mandate to deter attacks and protect civilians from human rights violations by the Janjaweed and government forces. The Darfur Peace Agreement (DPA) of May 5, 2006, paved the way for the deployment of a better equipped UN force to augment African Union Forces in a so-called "hybrid force," but the Khartoum government refused to allow UN troops into Darfur despite the fact that 10,000 UN forces were in southern Sudan. UN Security Council resolution 1706, which Khartoum defiantly rejected, provides a force of 22,500 troops and civilian police.

Sudan continued to criticize the international media for exaggerating the conflict in Darfur and portrayed the quest to deploy UN forces as a Western plot to undermine Islam and threatened to fight any UN force in Darfur "as Hezbollah fought the Israeli forces."[31] The government is identified as the perpetuator of the genocidal destruction of the Darfur civilians, and the UN resolutions underscored this claim.[32] The UN resolution 1556 in 2004 demanding the disarmament of the Janjaweed by the Sudanese government was not honored.[33] Resolution 1556 permitted the prosecution of Sudanese government officials by the International Criminal Court.[34] Resolution 1593 underscored the reality that the Sudanese government was not a worthy partner to deal with regarding war crimes and crimes against humanity.[35] Khartoum's complicity in the killings has met with condemnations from the global community. For its complicity in the genocide, the African Union has denied Sudan the chairmanship of the organization for the second time. Prime Minister Tony Blair observed that the Sudanese president was not complying with international law and was "playing the international community," so President G. W. Bush called for UN sanctions on Sudan.[36] Amnesty International, the International Crisis Group and Human Rights Watch, jointly called on the Security Council "to fulfill its responsibility to protect and the deployment of a UN robust peacekeeping force to protect Darfur civilians."[37] Despite the United Nations' refusal to classify the Darfur killings as genocide, the deliberate and selected killings of the ethnic black minorities, including children, and raping of women rendering them social outcasts, fall within the purview of the United Nations, which has both legal and moral obligations to protect the afflicted minorities. The Khartoum government, however, continued to taunt and bar the United Nations from deploying a robust mission with a protective mandate in Darfur, under the cloak of the UN primary principle of sovereignty and the related principle of domestic jurisdiction to prevent the United Nations from honoring its protective obligation in Darfur. On Tuesday, June 12, 2007, Sudan accepted the so-called "hybrid" UN–AU peacekeeping force to be sent to Darfur following a US–Chinese compromise. Under the new plan, the AU would run the day-today operations, while the United Nations was expected to have an overall control of the forces. However, Sudan strategically impeded its immediate deployment.[38]

The crisis was commonly described by the United Nations and Western states as "the most serious humanitarian emergency in the world today."[39] The United States and NGOs, such as Physicians for Human Rights and Justice Africa, called the crisis genocide.[40] The UN Security Council, the EU, and a variety of NGOs (Amnesty International, Human Rights Watch) all acknowledged the complicity of the Sudanese government in large-scale crimes against humanity and ethnic cleansing in Darfur.[41] Notwithstanding this realization and the subsequent indictment of the Sudanese President by the ICC for war crimes and crimes against humanity, the global community ironically refused to consider intervention in Darfur to protect the civilians. The specter of crimes against humanity and various forms of violations in Darfur by the government and its proxy militia sustained by ideologies of "ethnic cleansing"; territorial expansionism and political dictatorship continue to haunt and shock the conscience of humankind today, yet we have failed to give an appropriate decisive robust interventionist response to save the poor Darfurians. The next section examines why countries, specifically the UN Security Council, the United States, and its EU allies and Africa, denied Darfurians protection despite the overwhelming evidence of atrocities that meet the threshold of and global demand for robust intervention in Darfur.

International Responses to Darfur Killings

Darfur killings were repeatedly described as representing a supreme humanitarian emergency that met the threshold of robust intervention. Despite a recognition of the importance of human security and an increased focus on civilian protection, UN Member States that overwhelmingly endorsed the international norm of responsibility to protect failed to intervene robustly to save Darfurians from atrocities in defiance of global demand for it. In this section I explore the factors that dissuaded the Security Council, the United States, and its EU allies and Africa from intervening despite the overwhelming evidence of atrocities. What remains clear is that the international community's unwillingness to take robust "responsibility-based" action in Darfur is the result of a plethora of neorealist considerations of Member States.

United Nations Security Council and Darfur

It is a puzzle that despite the promise of "never again" following its failure in Rwanda and Member States' overwhelming support for responsibility to protect, the Security Council failed to muster the political will necessary to act decisively to end the humanitarian catastrophe in Darfur. During discussions on resolutions 1556 (2004) and 1564 (2004) on Darfur, Security Council members repeatedly emphasized Sudan's responsibility as well as the international community's collective responsibility to protect Darfurians in the absence of Sudan's success[42] and yet they refused to deploy a robust mission in Darfur in the absence of Sudan's consent. The Security Council's failure to intervene robustly in Darfur was obviously influenced by realpolitik in countries, particularly China and Russia, that have political and economic interests in Sudan.

The United Nations' first action on Darfur took place on November 7, 2003, when the UN Office for the Coordination of Humanitarian Affairs (OCHA) warned that Darfur had become the worst humanitarian crisis since 1988.[43] While the atrocities worsened, the UNSC prioritized the Naivasha negotiations in Kenya to end the North–South Sudanese civil war, hoping that its settlement would automatically douse the Darfur war. This false optimism caused the Security Council to discount the warning signs of a growing crisis in Darfur. The Council kept the Darfur crisis off its agenda, fearing that its discussion would scare the Sudanese government in Khartoum to pull out of the Naivasha talks.[44] As a result, Security Council resolution 1547 (2004), which established the United Nations Advance Team in Sudan to prepare for a future UN operation following the signing of the Comprehensive Peace Agreement, barely mentioned Darfur. However, it gradually became impossible to ignore the evidence of massive human rights violations and grave humanitarian needs in Darfur. After repeated appeals by humanitarian and human rights groups, coupled with Secretary-General Annan's and many European and US foreign ministers' visits to Darfur, the Security Council first addressed the Darfur crisis with the passage of resolution 1556, which characterized the crisis as a threat to international peace and security. This resolution condemned the abuses and demanded that the Sudanese government disarm the Janjaweed and hold human rights abusers accountable. The resolution threatened economic sanctions if Sudan failed to do so and imposed a ban on the sale of arms to all non-state actors—rebels and Janjaweed militias, but not the Sudanese government. It also failed to mention the complicity of the Sudanese government in the abuses. It endorsed the AU's protection force, African Mission in Sudan (AMIS), which was mandated by the African Union's Peace and Security Council to monitor the April 2004 ceasefire agreement between the government and the rebels. It has been suggested that the SC's support for AMIS provided a convenient excuse to avoid direct involvement in Darfur.[45] Despite the SC's threats, the humanitarian crisis in Darfur worsened and the government failed to protect civilians, disarm the Janjaweed, and hold violators accountable.

In September 2004, the Security Council passed resolution 1564, declaring its "grave concern" about the Sudanese government's failure to fully honor its obligations. The Council reiterated its call for the government "to end the climate of impunity in Darfur" by holding those responsible for the widespread human rights abuse accountable. It called for an expansion of the AU's monitoring mission in Darfur and established the International Commission of Inquiry on Darfur to investigate the violence and to determine whether or not acts of genocide had occurred. It threatened the government with additional targeted sanctions such as "actions to affect Sudan's petroleum sector and the Government of Sudan or individual members of the Government of Sudan"[46] if it failed to comply with resolutions 1556 and 1564. The commission found that the government of Sudan and militias acted together in committing widespread atrocities in Darfur that should have been prosecuted by an international war crimes tribunal, but the violent acts did not amount to genocide. Confronted with these findings and repeated provocations from the Sudanese government, including its utter failure to disarm the Janjaweed and prosecute violators, the UNSC passed a considerable number of resolutions[47] based on which a number of targeted sanctions were issued.[48] It also passed resolution 1591, which established the United Nations Panel of Experts concerning Sudan with the mandate, among others, to monitor the Arms

Embargo on Darfur.[49] The Security Council then referred the Darfur crisis to the International Criminal Court (ICC) for investigation.[50] The ICC started its investigation and prosecution of individuals who "jointly committed crimes against the civilian population in Darfur"[51] in February 2007, despite much obstruction from the Sudanese government. The ICC Prosecutor, Luis Moreno Ocampo, in his report to the Security Council in June 2006, detailed his team's collected evidence of mass violations of human rights and international humanitarian laws. However, despite this high level of activity, the UNSC failed to address the key issue of deploying a robust UN mission into Darfur to protect Darfurians under siege by their own government. As the violence in Darfur continued, despite the signing of the Darfur Peace Agreement on May 5, 2006, Security Council action was still largely limited to investigations, inconsequential threats, and support for an AU force.[52] As the crisis worsened, the international community called AMIS a "failure," and the United States, influenced by its interventionist lobby, immediately argued for a UN takeover.[53]

On August 31, 2006, the UN Security Council adopted resolution 1706 "inviting" Khartoum to consent to a UN takeover.[54] The resolution demanded a rapid deployment of UN peacekeepers to Darfur and authorized the mission "to use all necessary means [. . .] to protect civilians under threat of physical violence [and . . .] to prevent attacks and threats against civilians." Most importantly, the resolution explicitly endorsed the concept of the R2P and its application to the case of Darfur by reaffirming the provisions of resolution 1674 (2006) on the protection of civilians in an armed conflict. However, the reference to R2P infused uncertainty into the implementation of the resolution. The Sudanese government objected to the deployment of UN forces in Sudan, declaring that the resolution amounted to "unjustifiable hostility against Sudan."[55] Al-Bashir made a restrictionist argument that Western states were abusing humanitarian justifications to legitimize neocolonial interference in the affairs of the weak.[56] Al-Bashir maintained that the details "actually confirmed our suspicion because the content of the resolution places Sudan under international trusteeship of the United Nations."[57] The Sudanese government insisted that it would not allow an Iraq-style occupation or foreign interference with President Bashir announcing "there will not be any international military intervention in Darfur as long as I am in power."[58] He also raised the specter of an international jihad against any such UN force.[59] The Sudanese position gained currency among some members of the United Nations during the discussion and voting on the resolution. China and Qatar abstained from voting on resolution 1706 (2006) because the Sudanese government had not yet consented to the deployment. Russia abstained out of an objection to the use of R2P language. Even supporters of the resolution remained unwilling to deploy UN forces without state consent.[60] Secretary-General Annan made this clear when he said "The fact is, without the consent of the Sudanese government, we are not going to be able to put in the troops. So what we need is to convince the Sudanese government to bend and change its attitude and allow us to go in."[61] In future, such an intervention would possibly be vetoed by China and perhaps Russia in the Security Council.

However, following a US–Chinese compromise, the Sudanese government in a meeting with the United Nations on November 16, 2006, in Addis Ababa, agreed to the creation of a hybrid AU–UN force on condition that the hybrid force would be predominantly African in character,[62] which would integrate the 7,000 already deployed AU forces. UNAMID was established on July 31, 2007, by UNSC resolution 1769, which authorized it under Chapter VII to use force to protect civilians and aid workers. UNAMID has, however, not offered meaningful protection for the Darfurian population.

The mission and logistics of the force represent a watered-down version of what was originally proposed. As a result of political maneuverings by China on behalf of Khartoum, the language of the resolution was modified giving the Sudanese government wide discretion in a variety of political and logistical matters that undermined full implementation of UNAMID's protective mandate. For example, the UN Secretariat generally awaited permission from Sudanese authorities on decisions regarding troop composition, permission for night flights, and land use.[63] The mission's purpose and mandate are vague and have been contested.[64] During discussion on resolution 1769, the role of civilian protection was heavily debated.[65] For example, a Sudanese defense ministry official, General Majzoub Rahamah, claimed that UNAMID did not have the right to protect civilians, that it only had the right to protect themselves.[66] The Sudanese government obstructed UNAMID's efforts by mandating that no non-African troops, except those from China and Pakistan, could participate in the mission. It denied even these troops port access, landing rights, flight restrictions, and monitoring and reporting the legal status of the force. The regime heightened the obstructionism to the extreme when Sudanese government forces attacked UNAMID convoys repeatedly throughout the first quarter of 2008. On July 8, 2008, the attack by armed militias on a UNAMID convoy in northern Darfur, in which seven peacekeepers were killed and 19 wounded, underscored the gravity and urgency of the situation.[67] However, in February 2008, UNAMID and the government signed a new Status of Force Agreement (SOFA) that established a legal framework for UNAMID's operation. This agreement was to address issues such as communications, freedom of movement, safety and security of troops, privileges, immunities, and logistics of entry and exit to and from war zones. Khartoum maintained its preference for exclusively African troops, but conceded to consider deployment of non-African forces in consultation with UNAMID on a case by case basis. Despite this, UNAMID forces still face numerous obstacles in their attempts to fulfill their protective mandate and have not been able to offer meaningful protection for the Darfurian population, who have accordingly lost faith in the international community's will or ability to help them.[68]

A plethora of limitations undergirded by neorealist national interests explain the SC's failure to deploy a robust mission to protect Darfurians. Within the SC, divisions over the cause and nature of the conflict, as well as the culpability or responsibilities of atrocities, stymied efforts to deploy a robust mission in Darfur. Only the United States identified the Sudanese government as the perpetrator of deliberate attacks against civilians, and therefore responsible for human rights violations.[69] The other members held multiple parties responsible for the continuing violence. France, the Russian Federation, and the United Kingdom along with six nonpermanent members of the SC identified the armed rebellion by the SLA and JEM as the cause of the government-sponsored violence by the Janjaweed militia and state military forces. While the United Kingdom acknowledged the Sudanese government's failure to protect the citizens; it held the rebels equally responsible for the crisis and asked both parties to meet their obligations as laid out in the Humanitarian Ceasefire Agreement and relevant SC resolutions.[70] This lack of unanimity was reflected in the vote on resolution 1591 (2005), which enacted a series of sanctions against specified members of the Government of Sudan and the rebel movements. The Russian Federation abstained from these, indicating that while both sides were guilty, the resolution disproportionately punished the Sudanese government.[71] There were also disagreements on whether to label the Darfur atrocities genocide or not. With the exception of the United States, powerful international actors and the African Union chose not to use the word "genocide" because that would trigger the legal obligation for a forceful response. The divisions and disagreements handed UNAMID a vague watered-down mandate, especially with regard to civilian protection, in an attempt to mask the disagreements among UN member states on this issue.

The principle of sovereignty also trumped the concerns with the human rights violations in Darfur. In spite of the growing declared support for responsibility to protect civilians and numerous UN reports suggesting that state sovereignty is no longer a shield for human rights abuses, the Security Council was not willing to consider intervention in Darfur without the consent of the Sudanese government. The Council members and their resolutions on Darfur affirmed the sovereignty, unity, territorial integrity, and political independence of the Government of Sudan. China insisted that the international community should only play a secondary role to the Sudanese government in addressing the Darfur crisis, arguing that the sovereign government maintained primary responsibility for addressing the crisis.[72] There international response to the conflict in Darfur further illustrates that there is no consistent political will from powerful states, especially those that have declared their support for R2P, to commit military forces in ways that would challenge the traditional meaning of sovereignty.[73] Disunity among Council Members and the Secretariat informed by their strategic national interests undermined the clarity and unity of purpose required to garner the political will to authorize a robust mission in Darfur.

United States and Darfur

Candidate George W. Bush was opposed to humanitarian intervention to prevent mass atrocities.[74] Yet early in his first presidential term in 2001, when he received a National Security Council memorandum outlining US inaction in the Rwandan genocide, he famously wrote "Not on my watch" in its margin.[75] By doing this, the president made a rhetorical commitment to avoid the mistakes made during the Rwandan genocide, and asserted the "Responsibility to Protect" civilians. This pledge, however, did not translate into meaningful action to intervene robustly in Sudan and protect Darfurians from blatant human rights violations. Despite the fact that the United States and Britain were the most vocal critics of the Sudanese government, neither government pressured China to stand down from its threat to veto strong Security Council action to halt continuing human rights abuses in Darfur.[76] The administration feared it would jeopardize its counterterrorism cooperation with the Sudanese government if it pressured the government to stop the Darfur conflict. Perceptions that the Darfur crisis was being ignored generated unprecedented congressional pressure and citizen activism across the United States. Congress unanimously declared that the Darfur violence was "genocide."[77] This compelled Secretary of State Colin Powell to hire an NGO, the Coalition for International Justice (CIJ), to conduct a survey and enable him to make a determination. The report affirmed that the Darfur crisis constituted genocide.[78] The finding compelled Powell to appear before a US Senate committee and declare that "genocide" was occurring in Darfur. President Bush made the same claim in an address to the United Nations.[79] However, despite the genocide declarations, the United States ruled out any urgent response to the crisis. Powell declared that the formal genocide declaration would not lead to concrete policy changes on the part of the US government.[80] President Bush, addressing the UN General Assembly, reiterated the US position that genocide was occurring in Darfur, but he did not urge new international action to address it.[81] The United States also ruled out the deployment of US troops in Darfur. Colin Powell told reporters at the State Department on May 5, "There is no army that is going to go in there and put down the insurrection. We have got to use the pressure of the international community on Khartoum."[82] Instead, Powell sent the issue to the UN Security Council, which in turn established a commission to determine if, indeed, genocide was occurring. The administration had other "more pressing priorities" than crimes against humanity in Darfur.[83]

As the death toll continued to mount, the United States remained the only SC member that has publicly acknowledged that the Darfur atrocities constituted genocide. However, this claim did not galvanize official US action to coordinate sufficient and strong international action to stop the violence. Top US administration officials claimed that the United States was doing everything possible to stop the genocide but this was not followed by concrete actions to improve the security in Darfur. The United States made some diplomatic investment in the Darfur peace process to promote a long-term solution. In fact, the signing of the Darfur Peace Agreement in May 2006 was heavily promoted by the United States. It has also provided significant financial support for humanitarian efforts to mitigate the humanitarian crisis.[84] Yet, it failed to articulate or pursue a strong action on the most immediate priority of stopping the violence and protecting Darfurians from killings. The United States, however, introduced or supported numerous UN Security Council resolutions on Darfur, condemning the violence and urging an end to atrocities. It even imposed some limited sanctions on those perpetrating war crimes and crimes against humanity. It supported SC resolution 1556 (2004) that endorsed AMIS, mandated by the African Union's Peace and Security Council to monitor the April 2004 ceasefire agreement between the government and the rebels. The United States provided financial and logistical support to the AU. It offered transport planes to bring AU troops to Darfur. It also coordinated with NATO members to provide planning and intelligence support to the AU. But an assertive international leadership for an international robust protective intervention would have stopped the atrocities. And yet the US and the international community's responses were restricted to rhetorical commitments to protect along with humanitarian assistance. It has been suggested that the SC's support for AMIS provided a convenient excuse to avoid direct involvement in Darfur.[85] As the crisis worsened in 2006, the international community called AMIS a "failure," and an international consensus emerged on the need for a UN peacekeeping force to stop the violence and restore security. The United States immediately argued for a UN takeover in 2006,[86] and the Security Council adopted resolution 1706 on August 31, 2006, "inviting" Khartoum to consent to a UN takeover.[87] The Sudanese government objected to the deployment of UN forces in Sudan, declaring that the resolution amounted to "unjustifiable hostility against Sudan."[88] However, the United States and China reached a compromise that compelled the Sudanese government to agree to the deployment of a hybrid force, UNAMID, that is predominantly African in character.[89] Despite being authorized under Chapter VII to use force to protect civilians, UNAMID has not offered meaningful protection to the Darfurian population because the US–Chinese compromise was a watered-down version of what was originally proposed. The mission's purpose and mandate are vague,[90] giving the Sudanese government wide discretion in a variety of political and logistical matters that undermine full implementation of UNAMID's protective mandate. The Bush administration claimed it was a leader on Darfur and was committed to "Responsibility to Protect" civilians from crimes against humanity such as genocide.[91] The administration's support for the R2P and President Bush's "Not on my watch" pledge did not manifest in appropriate robust action to deal with the Darfur crisis. More powerful factors and competing priorities apparently negated these potential motivators for necessary US action to stem the bloodshed in Darfur. As Paul Williams and Alex Bellamy best synthesize, three main obstacles account for the United States' tepid response to the Darfur killings.[92]

The first explanatory factor is that when the Darfur crisis started, the United States and other third parties were actively promoting the Naivasha Peace Agreement to end the seemingly long, interminable North–South Sudanese civil war. They feared that robust intervention in Darfur would jeopardize the agreement.[93] Western leaders were concerned about the negative ramifications for the North–South war. US Secretary of State Colin Powell affirmed this concern when he stated, "here is a concern that we don't want to put so much pressure on the Sudanese government, which causes internal problems that might make the situation worse."[94] A more important factor was the US–Sudan intelligence-sharing relationship on terrorism that diminished the US willingness to intervene robustly in Darfur. The United States had an isolationist foreign policy toward Sudan in the 1990s but since 9/11 it has pressured the Sudanese government to be "with" rather than "against" it.[95] Sudan subsequently welcomed the US offer and invited an intelligence-sharing relationship with the United States on the "War on Terrorism."[96] This war undoubtedly was of more priority to the United States than Darfur, and the Bush administration did not want to jeopardize Sudan's cooperation with it.

Accusations of the West using humanitarian justifications to mask their neo-imperial ambitions[97] sapped their enthusiasm and political will to intervene in Darfur. Many states, particularly in the Islamic world, have become increasingly suspicious of the West's professed humanitarianism after its post-9/11 interventions in Afghanistan and Iraq. The Sudanese representative articulated Sudan's skepticism of the SC when he asked whether the Council's "lofty humanitarian objective" in Darfur was a "Trojan horse . . . embraced by other people who are advocating a different agenda."[98] The argument has been that the West's professed humanitarianism is a façade for its neo-imperial ambitions of obtaining oil, the war on terrorism,[99] and destroying Islamic values. This argument gained currency by actions of noninterventionist non-Western states in the Security Council, such as China. As a result, the United States could not garner support of members of the SC because the Bush administration was waging a globally unpopular war in Iraq without a UN mandate. Graphic images of US soldiers abusing Iraqi prisoners in Abu Ghraib further heightened their skepticism about the US humanitarian agenda, and inevitably the noninterventionist states rejected the US call for humanitarian intervention in Darfur. Throughout its interaction with the United States on Darfur, the Sudanese government portrayed the US interventionist position on Darfur as part of a global American assault on Islam, and Al Bashir promised an international jihad against any third-party intervener. The United States and allies feared that the alleged Western plots against Muslims could whip up anti-Western sentiments and terrorist attacks by radicals[100] following the intervention by the West in Darfur. US intelligence officials even suspected that Saudi Arabian terrorist groups were training in northeast Sudan.[101] Western states had to opt for an appropriate response that would not ignite Islamic radicalism and make Sudan a haven for anti-Western terrorist groups as it had been in the early 1990s.

The Bush administration had clear leverage and opportunities to push for deployment of a robust peacekeeping mission in Darfur to halt war crimes and crimes against humanity there. The Bush administration, however, retained its international leverage and its political will for other concerns in the Middle East and in the larger "War on Terror." Consideration of geostrategic interests and foreign policy priorities militated against robust intervention and the people of Darfur lost out and paid the ultimate price of US inaction.[102]

Africa's Response to Darfur Killings

The apparent paralysis of the United Nations and the reluctance of the West to take active robust intervention compelled the African Union to assume the primary responsibility of halting the Darfur crisis. However, despite clear evidence of the Sudanese government's complicity in the atrocities and the AU's adoption of a more interventionist operational paradigm in Article 4(h) of its Constitutive Act, the AU refused to act without the consent of the Sudanese government. Article 4(h) of African Union's Constitutive Charter permits collective intervention in "grave circumstances, namely war crimes, genocide and crimes against humanity."[103] The AU's strict adherence to the cardinal UN peacekeeping principle of consent, coupled with other considerations, undermined its political will to implement its interventionist security regime articulated in Article 4(h) of its Constitutive Charter. The AU initially opted for a negotiated settlement of the Darfur conflict. AU's mediation efforts resulted in the first ceasefire agreement in September 2003. However, the security situation deteriorated and the African Union, with UN Security Council endorsement, deployed a small number of civilian observers in Darfur to monitor the June 9, 2004, ceasefire agreement between government and the rebels. This mission eventually evolved into the African Union Mission in Sudan (AMIS), but its operational efficacy was hindered by some limitations. A critical limitation of AMIS was the intervention debate that crystallized around the question of whether it had the responsibility to protect Darfurians or not. While Rwandan president Kagame insisted it did, and the AU indicated in a communiqué to the Security Council that its troops would use force to protect civilians if necessary,[104] some AU members expressed reservations.[105] Nigerian President Obasanjo, for example, insisted that his forces would only protect AU observers and operate with the consent of the Sudanese government.[106] Sudan rejected AMIS' broad civilian protection mandate. A compromise was reached that permitted AMIS to protect vulnerable civilians only in their immediate vicinity with the understanding that the Sudanese government has the responsibility of protecting the civilian population.[107] The Sudanese government frustrated AMIS repeatedly, insisting that AMIS troops were monitors, not peacekeepers.[108] Its operational capacity hinged on the consent of the Sudanese government that was also complicit in the atrocities. The AMIS mission also lacked the requisite financial and logistical resources and manpower. Despite the challenges, the AU persistently championed peace talks among the conflict parties, culminating in the Darfur Peace Agreement (DPA) signed on May 6, 2006, between the government and the SLA led by Mini Minnawi. However, the DPA was not supported by the JEM and the other SLA faction led by Abdoul Wahid. These non-signatories viewed the AU as biased in view of its active involvement in negotiating the agreement. This led to increased armed attacks on AMIS personnel by these spoilers hindering the mission's operational efficacy. The AU threatened non-signatories to either sign up or face sanctions,[109] but the groups failed to comply, compelling the Peace and Security Council (PSC) to impose targeted sanctions such as travel bans and frozen assets of the leadership.[110] However, the sanctions were largely symbolic since the AU lacked the political leverage and mechanism to enforce them. Ultimately, AMIS was unable to effectively contain the violence and protect Darfurians under threat. It became apparent that a more sizable, better equipped UN peacekeeping force was required, but that was opposed by the Sudanese government. The AMIS mandate was extended repeatedly throughout 2006, despite falling short of its protective function due to the recalcitrance of the Sudanese government.

However, the growing insecurity, a lack of cooperation from the government and other conflict parties, and the absence of peace continued to compel the AU to propose a transition to a more robust UN mission.[111] The PSC called for "... a transition from AMIS to a UN operation, within the framework of the partnership between the AU and the United Nations."[112] The PSC, however, stressed that the transition would require the consent of the Government of Sudan and on the understanding that the composition and leadership of the UN mission would be mainly African in character.[113] President Bashir rejected the proposal challenging the AU's authority to hand over its mission to another organization.[114]

The UNSC welcomed the PSC's decision and requested the Secretary-General to design a contingency plan to explore options for transitioning to a UN operation.[115] The UNSC visited Sudan and held high-level consultations with members of the Government of National Unity including President Bashir and senior leadership of the AU and AMIS to enhance the proposed transition.[116] On August 31, 2006, the UN Security Council adopted a robust resolution, resolution 1706, "inviting" Khartoum to consent to a UN takeover.[117] The resolution demanded the rapid deployment of UN peacekeepers to Darfur and authorized the mission "to use all necessary means [...] to protect civilians under threat of physical violence [and...] to prevent attacks and threats against civilians."[118] President Bashir reiterated his objection, particularly to the deployment of a Chapter VII UN mission, which he believed would amount to a recolonization of Sudan.[119] He cited resolution 1679 (2006),[120] adopted under Chapter VII, as an example of bad faith by the international community.[121] The Sudanese government's persistent rejection of an exclusive UN force led to a US–Chinese compromise that resulted in the creation of a hybrid UN–AU mission, UNAMID, with a watered-down mandate and logistics required for a robust mission. The Sudanese government agreed in principle on condition that the hybrid force would be predominantly African in character,[122] operating largely with the consent of the host government. UNAMID was established on July 31, 2007, by UNSC resolution 1769, which authorized it under Chapter VII to use force to protect civilians and aid workers. UNAMID has, however, not offered meaningful protection for the Darfurians population due to certain inherent pivotal factors. Despite the AU's adoption of an interventionist charter and the admission that the deaths of thousands of civilians required its intervention, it failed to adopt the appropriate robust operational mechanism to stop the killings. The AU's Constitutive Act articulates a strong commitment to human rights protection, but chronic obstacles to more effective action remain.

Though the AU has adopted a more interventionist operational paradigm that permits collective intervention in "grave circumstances," its tepid response to Darfur atrocities arguably suggests that the principle of noninterference continues to override human rights concerns in Africa. Traditionally, African leaders, informed by struggles against neocolonial subjugation, are predominantly committed to colonial-tinged, state-centric sovereignty and noninterference to safeguard their hold on power. As a result, a dramatic shift from the favored traditional state-centric philosophy to an interventionist's paradigm to protect fundamental human rights is undoubtedly challenged by protectionists of the old order. This explains why the AU insisted on the consent of the Sudanese government before deploying forces in Darfur. Closely linked to this factor is the fact that the original Organization of African Unity (OAU), unlike the United Nations, was leader-centered rather than grassroots or people-centered. The OAU as a regional collective security system was a creation from the top rather than a genuine African peoples' aspiration. This is captured in its preamble, which states, "We, the Heads of African and Malagasy States and Governments assembled in the City of Addis Ababa, Ethiopia . . ." On the contrary, in its preamble, the UN Charter articulates the will of the people with "We the *peoples* of the United Nations determined to save succeeding generations from the scourge of war . . ." The difference here is clearly discernible. While the UN Charter advanced the general will of the people of the world, the African Charter advanced and legitimatized the interests and aspirations of the leaders rather than the African people. Despite clear evidence of the Sudanese government's complicity in war crimes and crimes against humanity that met the threshold of its interventionist charter of Article 4(h), the AU refused to act without the consent of the Sudanese government. The AU operated in Darfur with the consent of the government of Sudan. The AU was reluctant to push too hard on the government, fearing it would result in noncooperation from and marginalization by the Sudanese government. In 2004, the AU issued a final communiqué after a "mini-summit" reaffirming the participants' commitment to Sudanese sovereignty and pledging the "rejection of any foreign intervention by any country, whatsoever."[123] The sovereignty and protection of the ruling elites was of greater priority than the protection of civilians.[124] African leaders have been, disparaged as a "club of dictators," mutually supporting each other against humanitarian intervention to end atrocities in Member States.[125] Obviously, the AU's adoption of an interventionist paradigm is mere rhetoric never to be realized as evident in Darfur because most AU members states are equally riddled with official massive human rights violations, corruption, and poor governance. They rationally fear that an intervention in another state with a similar situation to theirs would pave the way for an intervention in their own state and dislodge them from power as happened in most North African states, such as Gaddafi's Libya and Mubarak's Egypt in the current Arab Spring. An intervention in another Member State to protect human rights constitutes a "political suicide" to lose political power and none of them is prepared to sacrifice their political power and concomitant economic and social prestige for the sake of human rights of the ordinary people. African leaders as a result lack the political will and preparedness to generate effective robust interventions in Member States as evident in their pallid response to the Darfur crisis.

Although Article 4(h) legally empowers the AU to intervene in a Member State to prevent mass atrocities, the AU states disagreed on the mandate of the AU's mission. There were disagreements on whether AMIS had legitimacy to protect civilians under threat or not. As a result, Darfur was handed a weak AU mission with constrained mandate operating on the terms of the Sudanese government and failing to protect Darfurians. The AU's rejection of the ICC's indictment and warrant for arrest for President Bashir for committing war crimes and crimes against humanity[126] affirms its elitist, protectionist stance even in the face of overwhelming evidence of atrocities that meet the threshold of accountability of the ICC. Although the AU did not object to the indictment per se, it was against its poor timing. It feared that the arrest of the Sudanese president could worsen the Darfur crisis, despite the fact that two African members of the UNSC (Benin and Tanzania) had voted in support of the referral to the ICC.[127] The AU also feared reprisals against the peacekeepers by the Sudanese forces and the Janjaweed,[128] repeatedly requesting the UNSC to suspend the indictment by invoking Article 16 of the Rome Statute.[129] AU's noncooperation with the ICC resulted in a rift between its PSC and the UNSC and undermined their collective effort to deal with the crisis.

The AU also suffers from a chronic lack of operational capacity and institutional preparedness to deploy a coordinated, properly resourced, and legitimate international response to mass, atrocious crimes of the magnitude of Darfur killings.[130] The African Standby Force (ASF) that constitutes standby multidisciplinary contingents is established for rapid deployment pursuant to Article 4(h). However, it is not yet a potent regional force capable of sustained, intense combat operations. Most African militaries have weak advanced command, control, and intelligence. They have poor land, air, and naval capabilities in terms of adequate and appropriate weaponry, combat training, and strategic mobility. The AU also lacks a doctrine for responding to mass atrocities.[131] While emphasizing early warning, the AU lacks the necessary corresponding resources and capacity to prevent mass atrocities. The AU's financial and military resources are severely limited. AMIS, for example, lacked adequate soldiers at the early stages of its operation to fulfill even the authorized level of 7,000.[132] The problem is that only few wealthy African states bear the cost of the AU's regular and peacekeeping budgets, complemented by inadequate foreign support.[133] The multiple crises in Africa inevitably compete for the available limited troops and funds, derailing the AU's capacity to respond immediately to humanitarian tragedies, as the slow response to recent calls for AU peacekeepers in Somalia has demonstrated.

The AU intervened in Darfur without the requisite financing, logistic resources, and manpower. AMIS at its peak was a 7,000-strong force and was unable to protect all civilians under threat. The peacekeepers in Darfur were ill-prepared and insufficiently supported for the task of protection. The AU's incapacity became evident when it proposed to hand over control of Darfur to a better resourced and organized UN force.

Conclusion

The chapter explored why the strongest advocates of "responsibility to protect" failed to protect civilians suffering at the hands of their own government in Darfur. The crisis exposed the inherent contradictions and tensions between existing international principles and emerging concepts, such as, the Responsibility to Protect. While the crisis met or even surpassed some of the thresholds for invoking R2P, the international community dithered and failed to consider a robust intervention at the expense of thousands of the lives of Darfurians. Darfur exposed serious shortfalls in the UNSC, the USA/EU coalition and the AU's ability and willingness to use force to halt ethnic cleansing and protect civilians. The crises demonstrated that the international community's ambition to robustly protect civilians in violent conflicts is hampered by neorealist national interests of states that drive protective interventionist decisions. International community's pallid, tepid, divided, and slow response to Darfur killings has revealed that geostrategic interests trump the more humane or altruistic criteria in interventionist decisions to protect civilians from atrocities. Interventions to protect civilians are purely and politically motivated by the neorealist interests of states. Politicization of humanitarianism has therefore hijacked the world's moral duty of civilian protection. The pallid response by the UN Security Council and the US/EU coalition and the AU has raised doubts about the commitment of the global community to honor its obligation to protect civilians from catastrophic sufferings in violent wars. The response to the Darfur massacres has blighted the world's commitment to its post-Rwanda rhetorical promise that "never again" would it dither when heinous crimes are committed against civilians. The experience has exposed the political and operational challenges the international community faces in implementing its interventionist security regime and its responsibility to protect civilians from catastrophic sufferings in violent wars.

8

Moving Forward: UN Robust Peacekeeping and Civilian Protection in Civil Wars

This book has argued that robust peacekeeping works better than traditional peacekeeping in protecting civilians from killings, because by its designation, it constitutes a *formidable barrier* to human rights violations. The *formidable barrier* model posits that the success of UN peacekeeping missions in staving off deliberate civilian killings is a function of their deterrent capacity. Therefore, such missions' success is a direct function of their ability to constrain combatants' violent behavior, resulting in a cost–benefit calculation by the combatants that leads to the choice of peace over continued fighting and civilian killings. In this sense, the mission's capacity to raise the cost of continued fighting by combatants is dependent on the superiority of its capabilities over those of the combatants. A stronger mission force in terms of large force size, stronger firepower reflected in its robust rules of engagement, and stronger combative and repulsive capacities may alter the calculations of combatants and spoilers to choosing peace over continued fighting and killing of civilians. Robust peacekeeping, by designation, has the ability to manipulate the relative costs of continued fighting and civilian killings and the benefits of peace to have more preferable effects. Robust peacekeeping missions, characterized by greater force strength, use of heavy weapons, major power participation, rules of engagement to use "necessary means" to protect civilians as well as for self-defense, and a high cost-tolerance level can put up the necessary barrier or impediment to intentional civilian killings by combatants and spoilers. The underlying strategy of robust peacekeeping is to alter the parties' calculations so as to make it too costly to fight and more beneficial to choose peace. The basic presumption of the use of deterrent force to restore stability is that hostilities harden bargaining positions and attitudes rather than encourage concessions by parties who suffer costs. The probability of accepting a diplomatic resolution to a conflict by parties afflicted by human and other related costs in an ongoing hostile conflict is conditioned on the cessation of hostilities and attacks on conflict parties. A deterrent force can create stability encouraging political dialogue to resolve the conflict.

Findings

Quantitative analysis

The quantitative analysis in Chapter 5 supported the core argument that robust peacekeeping works better than traditional peacekeeping in reducing civilian killings. This finding provides support for the theory of force deterrence, and that a large, robust UN force deters spoilers from killing civilians. A unique finding is the contribution of large strength to mission success. This resonates with the Brahimi Report's recommendation for large troop size for mission success because large troops will be able to occupy large areas, effectively monitoring and implementing mission mandate and posing barriers of one sort or another to war-mongers.[1] However, the related notions of major power involvement and more militarily assertive missions or rules of engagement do not necessarily appear to restrain killings, and in fact are associated in some analyses with significantly higher civilian casualties. This is explained by the fact that major power involvement is tactically framed as imperialistic, and be used to rally the locals around the flag and defend their values against the aggressor or neocolonialist. The findings further inform us of the severity of identity conflicts and confirm the argument that highly culturally diverse peacekeeping forces can derail mandate implementation as in the UNAMSIL experience in Sierra Leone.[2] The finding that regional cultural similarity results in mission failures is striking since it defies the popular assumption by UN policy makers that regionalism is the better option for international security due to forces' similarity with neighbors. We also observe that borders matter, and that countries bordered by many states may be prone to increased civilian killings in civil war situations. It was also found that at least a proto-democratic status of the host state appears to be a harbinger of potentially higher deliberate civilian killing. Robust peacekeeping may be affected by late deployment, external interventions, the vastness of the host country, the number of borders, arms shipment and the number, the willingness to fight and the cost-tolerance level of the spoilers, as well as the mission force as exemplified by the robust mission (MONUC) in the Democratic Republic of Congo. In view of the findings and the need for more refined understanding of the process at work in pacification through peacekeeping, key representative cases of robust peacekeeping in Africa including the Democratic Republic of Congo and Sierra Leone compared with two cases of traditional peacekeeping in Rwanda and Angola, were examined to determine what might have accounted for the differences and ambiguities discovered in Chapter 5.

Case studies

The quantitative analyses presented in Chapter 5 indicated results contrary to expectation in terms of the impact of certain salient constituent elements of robust peacekeeping such as robust rules of engagement and major power participation, in lowering civilian killings. Due to various problems, such as coding and measurement decisions and the temporal order of variables, statistical analysis alone may not be able to capture the true impacts of such factors. A more detailed study of individual cases would help establish the actual factors that impacted civilian killings. A closer look at individual cases, therefore, serves to inform the author, as well as the reader, of other potential significant factors impacting on civilian killings. Robust peacekeeping per se may not be enough to stave off deliberate civilian killings and could be limited by certain operational and structural constraints. The effectiveness of robust peacekeeping could be hampered by late deployment, major power neorealist interests in conflict, external parties' or states' interventions or involvement in conflict, the vastness of the host country, the number of borders, arms shipments, the willingness to fight and the cost-tolerance levels of the spoilers as well as the mission force. In view of the quantitative findings and the need for a more refined understanding of the process at work in pacification through peacekeeping, cases of both traditional and robust peacekeeping missions are also comparatively examined to establish other potential significant factors, such as lack of major power interests, that impact civilian killings but not captured in the statistical models.

Contrary to the robust cases, the traditional cases characterized by limited force size, restricted rules of engagement that limited the use of force only in self-defense, use of small arms and without major power participation, did not constitute formidable barriers to prevent the killing of civilians by the spoilers. By their designation, the traditional cases lacked the firepower and offensive capability needed to alter the calculations of conflict parties by raising the costs of continued fighting and the benefits for the conflict parties. The case study supported the theoretical argument of the effect of large force, heavy arms, and robust rules of engagement and major power participation on lowering intentional civilian killings. The absence of these factors in our traditional peacekeeping cases seemed to have adversely affected the missions' capacity to stave off intentional civilian killings. The case studies captured relevant factors such as late deployment, mission force partiality and crimes, porous borders, poor topography and large country size, risk tolerance level of both mission forces and spoilers, UN bureaucratic and hierarchical structures and military command system, lack of conceptual clarity and agreements on the use of force, and external parties strategic interests in the conflict. These have adverse impact on the operational capacity of some missions in reducing intentional civilian killings. The findings thus have both theoretical and policy implications in the field of peacekeeping.

Implications for International Relations

In an era of collective efforts to stave off human rights violations, the United Nations needs to reexamine its guiding principles and its characteristics in order to address the hindrances to peacekeepers' contribution to the United Nations' pursuit of global peace and protection of civilians from human rights violations. Thus the United Nations needs to adopt a new operational paradigm to meet contemporary challenges posed by the changed, post–Cold War, violent civil war environment marked with horrendous human rights violations. Very few scholars or policy makers have discussed robust peacekeeping, but none has subjected it to empirical analysis as this book does. It assists in doing so by systematically analyzing the varied factors underlying the success and failure of missions, particularly in terms of human rights observance. This book differs from other works because most analysts focus on criteria such as mandate fulfillment, and the recalcitrance of the protagonists, to evaluate mission success. The missing link is that there is limited scholarly work on the relationship between UN robust peacekeeping missions and mission characteristics and intentional civilian killings. This book fills this gap because it is the first to systematically and empirically document the relationship between robust peacekeeping and civilian killings, using both statistical and case study models. This book focuses on civilian killings because this issue has gained primacy in discussions within the United Nations and global community. This book also focuses on immediate "success" when peacekeepers are deployed to complement the works of others, who emphasize long-term effects after termination of mission.[3] The focus on immediate effect is justified on the basis that long-term goals; for example, addressing the underlying causes of the conflicts can more successfully be pursued in an atmosphere of immediate secured peace achieved through robust peacekeeping. No purposeful and meaningful peacemaking can be successful without the immediate, initially secured enabling environment that robust peacekeeping provides. Borrowing from the cost–benefit theories of conflict termination, this book formulates the *formidable barrier* model of UN robust peacekeeping mission success using game theory to explain its effectiveness in protecting civilians from killings by spoilers.

Implications for Policy Makers

Suggestions by UN policy makers and peace practitioners to adopt robust peacekeeping to enable UN peacekeepers to protect vulnerable civilians in civil wars, signify that the research is topical and of interest not only to the academic community, but also to the informed public. This is especially true given the empirical findings that robust peacekeeping works better than traditional peacekeeping in lowering civilian killings when deployed in violent civil wars. This research conveys new and important information and sheds light on the operational capabilities of United Nations' robust missions to protect vulnerable civilians from deliberate killings by spoilers in violent civil wars. The findings identify certain policy gaps that need to be addressed if robust missions are to effectively implement their protective mandates.

The research finds that the United Nations lacks a functional military doctrine for peace operations and operational definition for robust peacekeeping. The Capstone Doctrine, recently developed by the Department of Peacekeeping Operations (DPKO) and Department of Field Support (DFS) to alleviate this gap,[4] does not seek to "override the national military doctrines of individual member states."[5] A more serious dilemma is the absence of conceptual clarity and articulation on clear guidelines on the use of force by mission forces to protect civilians. In the absence of a standard UN conceptual clarity, participating contingents in a mission rely on their domestic legal guidance. The invocation of national military doctrines imposes differing political constraints on contingents that may conflict with each other. Invocation of "national" caveats could lead to differential interpretation of the rules of engagement (ROE) as well as interference with command and control of missions in the field. Inconsistency in the rules of engagement can also complicate efforts to have a unified operational doctrine.[6] Discrepancies in operational doctrine on the use of force to prevent civilian casualties may result in differential responses to human rights violations against civilians by different contingents in the same operational theatre. The Security Council also lacks a coherent operational definition of "civilians' protection" and has used the terminology with multiple meanings in different mandates. This doctrinal void impedes the designation of appropriate protective mandates and planning and preparedness of mission forces to protect civilians under imminent threats. The absence of a protection doctrine may result in different interpretations by different contingents and stakeholders.[7]

The United Nations needs to develop a unified coherent doctrinal definition for robust peacekeeping and an operational doctrine on the use of force by robust missions. This should be done in consultation with troop contributing countries (TCCs) and other stakeholders such as non-governmental actors that operate in the same conflict theatre. Interagency coordination would mitigate potentially adverse effects of national or organizational doctrinal caveats on the mission's objectives. The new unified operational doctrine should establish principles and definitions, and the structure under which robust missions can use force to protect civilians under threat. It should provide coherent guidance articulating what "protection of civilians" means in operational terms. It must outline general elements of expectation and responsibility for the forces to protect civilians. The doctrine must underscore the fact that the very presence of peacekeepers in the face of threats to civilians should automatically trigger mission forces' obligations to protect civilians who are deliberately and brutally attacked by spoilers.[8] This will facilitate unitary interpretation of ROEs consistent with missions' protective mandates.

Most importantly, those general responsibilities should be factored into the comprehensive plan for robust missions, aiming at preventing and stopping systematic killings of civilian population using both soft and hard power tools and resources. The doctrine should guide the mission mandate and the pre-deployment training of mission forces. Prior to deployment, mission forces should be adequately trained in the protection of civilians from imminent threat of physical violence. Congruent understandings about the nature, basic principles and objectives of robust peacekeeping, civilian protection, use of force and rules of engagement would foster operational efficiency of the UN missions in executing their protective mandates.

Very critical to civilian protection is the United Nations' or a mission's capacity to collect and analyze appropriate intelligence to address threats to civilians. There is also the need to develop effective mission capacities to collect and analyze intelligence needed to either address immediate threats to civilians or predict potential crises that could rapidly escalate to violence.

The research also finds that the complex and divisive politics within the Security Council, fuelled by neorealist national interest, poses a hindrance to the deployment of robust missions. Council members' decisions to either deploy a robust mission or not are driven by their national interests and stake in the conflict. Altruism and the severity of atrocities are not the primary determinants to deploy a robust mission; rather, national interests are. The Rwandan genocide and Darfur killings are clear manifestations of the failures of political will and moral obligations due to divisive politics within the Security Council fueled by neorealist considerations.

To circumvent this obstacle in protecting needy civilians, the book recommends that whenever the SC is ineffective in making protective decisions to halt mass atrocities, the General Assembly's protective obligation should be invoked through the 1950 GA Uniting for Peace Resolution as it did in the 1950s on the SC's impasse over Korea.[9] Even though Article 24 of the UN Charter has entrusted the SC with the "primary responsibility for the maintenance of international peace and security," an exception is the 1950 General Assembly resolution 377, Uniting for Peace. This resolution guarantees the General Assembly (GA) the power to deploy military force when this is "necessary to maintain or restore international peace and security" when the Security Council, "because of lack of unanimity fails to exercise its primary responsibility for the maintenance of international peace and security."[10] Despite the Soviet Union's protest that this undermined Chapter VII of the Charter,[11] the International Court of Justice subsequently advised that the Assembly had "secondary" responsibility for peace and security, and if the use of veto prevented the Council from acting, the Assembly could organize peacekeeping operations at the request or with the consent of the States concerned.[12] Uniting for peace is therefore applicable when the SC is unable to unanimously sanction appropriate response to threats to international peace and stability, such as atrocities against civilians. The universal membership of the General Assembly gives it a comparative advantage over the Security Council in making well-informed protective decisions. The GA should play a very active role in emergency response following a SC's manifest failure to protect populations from genocide, war crimes, ethnic cleansing, and crimes against humanity.

The national interest of the members of the Security Council also influences the language of mandates for UN robust peacekeeping operations. As argued earlier in Chapter 3, SC mandates are intrinsically political and their language tends to be nuanced rather than clear. In most cases mandating resolutions are drafted in deliberate obscure terms. Mission mandates are sometimes vague, ambiguous, or compromised, and designed to satisfy SC members' interests to avoid gridlock and obtain the consent of all Council members. Usually the protection provisions of the mandates are worded with constraints such as "protect civilians under imminent threat and are in the immediate vicinity," "providing protection within the resources and capabilities of the force," "it being understood that the protection of the civilian population is the responsibility of the government," and "without prejudice to the responsibilities of the host government." The protection mandates are normally not given the highest priority within the overall UNSC resolution.[13] Mandate ambiguity may secure the consensus of powerbrokers necessary to get a resolution adopted, but, "Mandate ambiguity, coupled with fear of censure when mistakes are made can create a disincentive to innovative action to protect civilians."[14] Mandates couched in unclear language are susceptible to multiple interpretations by command and contingents and are difficult to translate into operational orders.[15] For example, UNAMIR's ROE permitted the use of all available means to halt ethnically or politically motivated criminal acts such as executions and attacks on displaced persons or refugees, and suggested that troops may be morally and legally required to act.[16] Force Commander Dellaire broadly interpreted this to mean an obligation and authority to use force to protect civilians from atrocities. However, the UN Secretariat took a more restrictive view of the mandate and insisted that UNAMIR was not to intervene.[17] It is obvious that mission forces cannot operate effectively in an environment of ambiguity.[18] Strong and specific protective mandates, backed by adequate resources, will stimulate capable responses to protect civilians. It is high time the General Assembly and the civil society advocated and pressured governments of Security Council members to provide robust missions with unambiguous mandates that clearly articulate that peacekeeping protective responsibilities.

The study finds that the lack of a common training doctrine for mission personnel undermines mission coherence, coordination, and command and control. The Capstone Doctrine does not "override the national military doctrines of individual members states"[19] depriving the United Nations of the unique opportunity to make a universal functional military training doctrine to guide the preparation of troops for deployment. Compared to NATO, the UNDPKO and the DFS suffer from a severe lack of military planning capacity,[20] which ultimately dampens the political will of Western nations to commit troops to UN peacekeeping operations. A robust mission in a dangerous and complex conflict environment would require a coherent and streamlined UN training doctrine similar to NATO's to guide the mission. The UN training doctrine should override national training doctrines since such diverse doctrines would ultimately result in policy and operational inconsistences and incoherence that would incapacitate the mission.

Closely linked to the lack of a unified training doctrine is the absence of a centralized pre-deployment training of forces about mission tasks, such as the use of force to protect civilians from imminent threat of physical violence. In the absence of a clear guidance from the Council, senior mission leaders and contingent commanders make decisions about mission strategy and tactics to protect civilians. However, each Member State has different training doctrine. The results are various inconsistent and diverse training doctrines among troop-contributing states that ultimately lead to discrepancies in operational doctrine on the use of force to protect civilians. The sequel is marked differences in the responses to human rights violations by different contingents in the same operational theatre. This substantially reduces the effectiveness of UN operations, especially in the most challenging and critical elements, such as the use of force to protect civilians in civil wars.

Specific capabilities are required for mission forces to protect civilians and certainly a unified training doctrine as well as pre-deployment joint training for all mission troops is relevant for mission success. These will help troops reach a consensus on the use of force to protect civilians. NATO's success in maintaining consensus during the Kosovo crisis is cited as exemplifying a collegial approach to successful conflict resolution involving different TCCs.[21] Pre-deployment training based on a unified training doctrine will boost skills in the field, build the necessary competencies, and *esprit de corps* enhance unanimity of purpose, enabling troops to successfully execute their protective mandates. Training would adequately prepare and provide troops with the requisite mental attitude much needed for robust missions in turbulent war theatres such as Darfur. This training will prepare troops psychologically to reduce the impact of incurring casualties and seeing some horrifying atrocities during operations.

This book recommends that the DPKO consult with TCCs about their training manuals and examine the caveats on the rules of engagement, training, weapon capabilities, and operational doctrines on use of force to protect civilians. The DPKO should harmonize the manuals into a unified training matrix to guide pre-deployment trainings to avoid conceptual confusions and differential application of use of force to protect civilians, as happened to the Uruguayan contingent in MONUC. The DPKO's standardized training doctrine should articulate clear operational requirements for protection-related tasks and be disseminated by DPKO to all participating TCCs. The DPKO integrated training team should coordinate with senior political and military leadership and visit force headquarters, staff colleges, and training schools of TCCs to explain the training modules.[22] In the operational theatre, the mission force headquarters must review newly arriving contingents to assess if they meet pre-deployment training requirements. This helps a command to determine their operational readiness to undertake its protective mandate and to address deficiency, if any exists. The DPKO can co-opt the G8+ and borrow from its Global Peace Operations Initiative (GPOI), credited with successful pre-training programs.[23]

The study finds late deployment of missions as an undermining factor in UN missions' capacity in protecting civilians, as occurred in the Rwandan genocide. The United Nations lacks a deployable pool of personnel ready for an immediate deployment. The Peace panel on "United Nations Peace Operations"[24] recommended the need for a rapid deployment capability with the objective of deploying missions within 30–90 days. Pursuant to this, the UNHQ proposed a Standby Arrangements System (SAS) for troops and civilian police to be rapidly deployed in response to crises to stave off escalation. The United Nations is in consultation with Member States seeking their commitment to support the SAS by contributing troops that can be accessible to the United Nations for rapid deployment. However, Member States have never mustered the necessary political will to make this happen. Strategies are still afoot to attract more countries' support.[25] Improving performance will require the United Nations to come to grips with the fact that its deployment policy is fundamentally unsuitable for rapid military and civilian deployment. The United Nations must consult with TCCs to address their deployment concerns or hiccups and harmoniously formulate a rapid deployment mechanism with their input for speedier deployment.

Closely linked to this is the requirement to reexamine procurement rules and regulations as they pertain to peacekeeping. UN peacekeeping operations are subject to the same rigid rules for procurement and contracting as the rest of the UN Secretariat. This requirement fundamentally delays early deployment of missions. Advanced procurement and budgetary authority procurement for the mission would enhance speedier deployment.[26]

The United Nations suffers from mandate–resource discrepancy and it acknowledges the problem of resources, particularly funding, personnel, logistics, and equipment since members have been reluctant in honoring their financial obligations. As a result, there is a growing disparity between the capacity of the United Nations and the demands of peace and security operations that require resources. The United Nations still struggles to scramble together enough logistics, funding, and material capacities to compose a credible mission in executing their protective mandates in current operational theatres. The UN peacekeeping operations have insufficient high-quality personnel, inadequate equipment, sub-standard communications systems, equipment, and other standard capacities that most well-developed militaries expect. [27]This reduces the ability to operate.

The Brahimi report acknowledges that troops need to be resourced before they can protect civilians under threat.[28] The DPKO is in consultation with Member States to store key mission equipment with long procurement and delivery time at the United Nations Logistics Base (UNLB) in Brindisi, Italy.[29] The United Nations must also make strategic reforms to ensure transparency and eliminate corruption and waste in the organization to encourage members to honor their financial obligations. For example, the United Nations' effort in Angola was undermined by waste and corruption. Tens of millions of dollars were wasted due to procurement fraud involving corrupt UN officials and suppliers, according to an April 28, 1995, report by the UN Under-Secretary for internal oversight.[30]

The findings show that major powers lack the political will to deploy robust missions and contribute troops to peacekeeping missions in conflicts of less strategic interest to them. Their strategic national interests undermine the clarity and unity of purpose required to garner the political will to authorize a robust mission. Consideration of geostrategic interests and foreign policy priorities militated against robust intervention by major powers to protect civilians in Rwanda and Darfur. Several of the world's most capable militaries are reluctant to accept casualties while fighting nonexistent threats or fighting in "unnecessary wars." They are involved in other commitments deemed pertinent to their national security interests. Their selective engagement in UN peace missions has gravely damaged the United Nations' reputation and efficacy in responding to global threats. The complexity of robust peacekeeping requires the participation of major powers that have the requisite military capabilities, particularly advanced weapons training, intelligence, mobility, and targeted firepower, to mitigate the risks of robust peacekeeping and effectively execute protective mandates.[31] The support of the SC demonstrates the global community's commitment to robust missions that enhance the missions' legitimacy and credibility. This signals to potential violators that the world would not stand by when heinous crimes were committed against innocent civilians and that human rights violations carry real costs. The support of the SC also boosts the world's confidence in the United Nations' capacity to protect civilians. Major powers need to switch from relying solely on national interests to using their conscience in decisions on global problems, especially in the post-9/11 era, where the global community faces common threats from weak states. Obviously, it would be naïve on the part of major powers to ignore the relevance of national interests in foreign policy discourse; however, it would be a strategic blunder to consider some conflicts as "non-existential threats" or "unnecessary wars" because they constitute the fundamental sources of "existential threats" to major powers. Leading scholars, such as Francis Fukuyama, have argued, "Since the end of the cold war, weak and failed states have arguably become the most single important problem for international order."[32] Susan Rice, former US Assistant Secretary of State for African Affairs, clearly confirmed this conviction when she testified that

> Much of Africa has become a veritable incubator for the foot soldiers of terrorism. Its poor, young, disaffected, unhealthy, uneducated populations often have no stake in government, no faith in the future, and harbor an easily exploitable discontent with the status-quo . . . these are the swamps we must drain . . . to do otherwise, we are going to place our security at further and more permanent risk[33]

Weak and failed states, due to their inherent weaknesses, are safe havens for terrorism and international criminal activities, such as drugs trafficking, money laundering, human trafficking, and piracy, all of which finance terrorism. Somalia is now a failed state and, like Sudan, it has become a den for terrorism and other criminal activities that threaten the strategic interests of major powers in Africa. The challenges the global community faces today are complex and demand collective efforts. Confronting these problems requires an operational paradigm shift by the major powers from primarily selective involvement to active involvement in resolving all conflicts in all regions. Selectivity and neglect of other conflicts would result in failure to overcome these challenges.[34] The international community demands members of the SC to coexist at least in mutual respect through diplomacy to help reach consensus on protective mandates and support and contribute toward robust missions to protect civilians from human rights violations.

The book finds that the UN missions' capacity to protect civilians is limited by severe constraints inherent in the UN system, particularly, the characteristic, heavy UN bureaucratic and hierarchical malaise of top-down structure with complex procedures, where every action requires approval from the top. This sometimes demobilizes the operational capacity of the field troops due to delays on operational decisions by DPKO. MONUC suffered this limitation and its forces were compelled to adopt a reactive posture rather than a proactive attitude.[35] The UN "failure to protect Rwandan citizens" is attributed to the general disorganization and bureaucratic chaos of its composition.[36] In one of General Dallaire's parting cables, he said that "the [UN] force has been prevented from having a modicum of self-respect and effectiveness on the ground."[37] When he warned the UN headquarters about the planned massacre and sought permission to intervene, the UN bureaucracy did not want to violate Rwanda's "sovereignty" and refused to act without the consent of government that masterminded the massacre. If Dallaire had been granted the flexibility as the commander on ground to seize the weapons and planners of the Rwanda massacre, the genocide could have been averted. It is high time the United Nations' hierarchical command and control structures became flexible, in view of the complexity of robust peacekeeping. This will enable field commanders on the ground to take certain tactical initiatives to confront day-to-day threats to civilians. In the fluid, volatile, and turbulent conflict environment of robust missions, commanders need to make on-the-spot or daily tactical decisions expedient to confront threats to civilians from conflict spoilers. Waiting on the UNHQ, located in New York and absent from operational theatres, for authorization before making tactical decisions results either in delays or in inappropriate tactical decisions that could hamper robust missions' capacity to protect civilians under threat. The structural flexibility should, however, not negate the relevance of the UNHQ's strategic guidance to robust missions. Information sharing among the field commanders, UN headquarters, and the Security Council should be strengthened to ensure unanimity of purpose and operational coherence.[38]

The findings show that UN peace missions lack an integrated chain of command and common procedures. Even though different national contingents are deployed under one Force Commander, the participatory preconditions allow each contingent to report to its contingent commander, who is first accountable to his own government, before obeying the Force Commander. These limitations constrain missions' operational capabilities in protecting civilians. While Western militaries recommend that the UN command and control should more closely resemble the more centralized operational command structures of the EU or NATO, other troop contributors are comfortable with the more decentralized model of the United Nations.[39] The proliferation of different command structures and "caveats," which risk gaps in strategic and tactical decisions, adversely impact the SC's and field commanders' authority and effective conduct of robust missions. This book recommends that the Security Council consult with TCCs and harmonize their caveats to develop a consensus so that all troop contributors, notwithstanding different military traditions, are comfortable with UN command and control arrangements.[40] The UNHQ should have control of robust missions in order to secure the authority of the Force Commander and ensure that contingents respect the United Nations and his command. However, this should not deny field commanders the flexibility to make tactical decisions to deal with threats to mission and civilians.

The findings reveal that crimes committed by the UN peacekeepers in areas of deployment affected successful achievement of their protective mandates. For example, MONUC forces' involvement in sex, bribery, corruption, and illicit trading scandals has undermined its credibility and reputation as a legitimate peace broker. As a result, some conflict parties are not cooperating with it. The assumed partiality of UN peacekeeping forces against some conflict parties undermines a mission's ability to successfully implement its mandate. In the DRC, MONUC's joint operations with the Congolese Armed forces (FARDC), discredited for massive human rights violation, against other conflict parties undermined its impartiality and legitimacy as an honest broker. As a result, the aggrieved parties attacked the UN mission and the civilians they were protecting. A UN mission must weigh the ramifications of partnering with conflict parties in the execution of mandates. In the DRC, the joint operation with the corrupt and brutish FARDC was a tactical error.

A closely related problem is that contributing states have disciplinary jurisdictions over their own troops, and neither the United Nations nor the host state can take legal action against offending troops. Troop-contributing states insist on retaining disciplinary jurisdiction over their military forces, wherever they are deployed. UN officials can at best send miscreant peacekeepers home, where they seldom face disciplinary measures. However, the United Nations, rather than the contributing state, has to shoulder the blame for the failure to appropriately punish guilty troops.

It is time for the United Nations to take the necessary corrective measures to "win the hearts and minds" of the international community, particularly the local populations, as an honest broker of peace. The United Nations has taken steps to investigate peacekeepers' crimes but the efforts have been woefully inadequate due to certain inherent contradictions. For example, the introduction of management training modules and the launch of a misconduct tracking system in mid-2008 to monitor and report mission forces' misconducts has yielded mixed results.[41] This book recommends that the oversight system should be complemented with a thorough examination of the issue of disciplinary jurisdiction of a peacekeeper. There is an expedient requirement to overhaul the United Nations' model Memorandum of Understanding (MOU), the template for the agreement it signs with each troop-contributing state that defines the terms of personnel deployment. The MOU is the tool for holding TCCs accountable for their peacekeeper's behavior. The revision must unequivocally articulate the TCCs' obligations regarding the conduct, discipline, jurisdiction, investigations, and accountability of crimes committed by mission forces. The United Nations' enforcement of its zero tolerance policy should not simply be limited to deportation of miscreant troops who go unpunished by the home governments, but should require the United Nations to have disciplinary jurisdiction over offenders to punish them, redeeming the waning reputation of the organization. The book recommends a new international convention or law be enacted to allow the United Nations and host states to have some level of disciplinary jurisdictions over troops that commit certain crimes that add grief and pain to families already experiencing turmoil. There is a need for more force abuse training before deployment to make them more sensitive and more attuned to the local and national cultures, manners, languages, and plights.[42]

The effective protection of missions requires UN partnership and coordination among all stakeholders of UN peacekeeping. The UNDPKO's Lesson Learned Unit (now renamed the Best Practices Unit) created in 1995 produced a report on the successes and failures of its missions that was criticized as less vigorous. The Units 1996 general reports entitled "Relations with Local Population" did not discuss civilian protection or crimes against humanity at all. The GA, in response to the Brahimi report, convened a Special Committee on Peacekeeping Operations, which accepted a number of recommendations, including measures for improving communications and coordination among stakeholders to ensure the quality of pre-mission planning and preparations and over all mission success. However, these recommendations have not been fully implemented. This book recommends that a Peacekeeping Council comprising experienced TCCs and practitioners be formed to coordinate resources, improve training and supervision, analyze past and ongoing peacekeeping missions, and submit report and recommendations to the United Nations Security Council for improved robust missions. This Council will be helpful by tapping their operational experiences to help shape appropriate policies and strategies to effectively protect civilians.[43] The Peacekeeping Council should assist the DPKO to develop appropriate protective doctrine, policy, and training to prepare troops for missions with Chapter VII and civilian protection mandates. A more flexible interaction within a key coalition of peacekeeping stakeholders, especially in the design and oversight of operations, will address caveats on the rules of engagement, training, weapon capabilities, and operational doctrine on the use of force to protect civilian population. This will enhance the United Nations' sufficient logistical and procurement capacities for rapid deployment, and obtain sufficient robust contingents to protect civilians.[44]

Implementation of the R2P will also foster the global community's commitment to protect civilians under threat. Despite the UN Member States' unanimous endorsement of R2P, the process of institutionalizing and making it operational has slowed down. The international community's tepid response to the Darfur crisis manifests the challenges R2P faces. Its status as a new "norm" of international conduct has been strongly resisted by some states,[45] because the concept is still understood differently, and sometimes only marginally, by various commentators and groups. Since 2005, there has been continuing disagreement over what R2P precisely entails and who bears the responsibility to act. Effective protection of civilian populations under threat, therefore, requires moving R2P from rhetoric to action.[46]

In his January 2009 report, Secretary-General Ban Ki-Moon articulated a four-component implementation package. Capacity building calls on the international community to help states meet their inherent responsibilities to protect their populations through confidential or public suasion, education and training, and other forms of material assistance. Closely related is the potential development of a civilian rapid deployment force that could help in the prevention of R2P crimes.[47] The report heavily focuses on prevention and the need to develop the capacity for early warning and assessment within the UN system. This will require a timely flow of accurate and relevant information on potential mass atrocities within a country and the institutional capacity to assess the information and understand its implications. The Secretary-General, in the report, suggested the consolidation of the analysis and sharing of information in a single office for the Prevention of Genocide and the Responsibility to Protect. The operationalization of R2P requires timely and decisive responses as well as effective collaboration with regional and subregional organizations to respond to an unfolding or impending mass atrocity. The report acknowledges the need to build regional capacity, learn from different regions' experiences and build consensus on civilian protection. The international community also needs to transform R2P from aspirational to a binding regulation by resolving it into rules of customary international law in order to effectively operationalize it and protect civilians. R2P today is simply a normative principle of political and moral, but not legal, weight. What is needed now is the generation of the necessary political will to back it up. This can be painstakingly built, through good information, good arguments, and good organization. It can be effectively done when the economically and militarily capable states, such as the United States, take the necessary steps to bolster and participate in UN actions.[48] UN Member States should also make prevention of mass atrocities an important national security interest and foreign policy objective. To that end states should develop tailored strategies to reduce a risk of atrocities, and the United Nations and other stakeholders must monitor states at risk of atrocities. The international community's response should not be just a humanitarian goal, but a legal responsibility that outlines the diplomatic and military steps the United Nations is prepared to take.[49] In a globalized world, civil societies and NGOs have commensurate responsibilities to promote the transformation of R2P from an aspirational goal to a reality by lobbying their governments and individuals and fully funding the effort of the unit for mass atrocities. They can promote the R2P through advocacy, education, research, information, and analysis on civilian protection.

Leadership matters and the international system require effective leadership to successfully confront the growing threats of human rights violations, particularly in civil wars. Nothing gets done in the system unless one state takes the initiative to lead. The United Nations becomes operationally potent when a state leads, but fails when nobody does. Leadership involves providing vision, direction, and an effective mobilization of resources to take an action.[50] The key to credible, implementable "robust" peacekeeping mandates hinges on effective global leadership.

The United States, the "only remaining superpower" in a unipolar world, with awesome military, economic, and cultural power, is the one state capable of exercising global leadership. The United States still sits atop the power hierarchy among major powers, not as a dominant power, but as the leading one.[51] It has unmatched capabilities and leverages in preempting or confronting international threats such as human rights violations, human security, terrorism, climate change, drug trafficking, and pandemics. Its economy, despite its current travails, is still the largest and most dynamic and still produces almost one-quarter of the world's gross output.[52] Its military, the only one among all militaries, possesses global reach. Diplomatically, the United States remains at the center of most international transactions.[53] Despite the recent spurt of anti-Americanism over the past decade, many states still look up to the United States for help and protection. Most states rely on the United States as a balancer of power in their regions, as well as their sole protector against potential regional and global threats from states and nonstate actors.[54]

The United States' overwhelming military and economic capabilities and global reach cast it as the only power capable of global action and leadership. As a result, most states still look up to the United States to provide the necessary leadership for solving or managing international problems. Unfortunately, the United States has fallen short of this expectation and remains rather indifferent to the United Nations, a proclivity evident in the Clinton administration's Presidential Decision Directive 25 (PDD 25). This precluded continued US participation in UN peacekeeping, and in wars that are of no strategic interest to it. George W Bush's administration's preference for unilateral action and ambivalence about multilateralism and distrust of the United Nations also explains why the United States failed to lead the United Nations.[55] Daniel Patrick Moynihan, former US ambassador to the United Nations, articulated the US distrust of the United Nations when he described the United Nations as a "dangerous place," and succeeding officials and legislators have preserved the perception.[56] US post-9/11 unilateralism in Iraq and Afghanistan, covert operations in Yemen, Pakistan, and Somalia, most without UN authorization, have been protested by some members of the United Nations. Although the US power qualifies it to lead the United Nations, its ambivalence about UN multilateralism renders it a reluctant leader.[57] Rather than leading the world organization, the United States is seen by other nations as an obstruction in the United Nations, and it sees the organization as an instrument of its foreign policy.[58] The United States is actively involved in UN activities when they serve its immediate, neorealist strategic interests. The United States thwarted UN deployment of robust mission in Rwanda during the genocide but intervened robustly, disregarding SC, in Kosovo, a conflict that pales in comparison to the Rwandan genocide. In the spring of 2003, the United States rejected the SC's authority, resulting in the unilateral decision to invade Iraq and topple the Saddam Hussein regime. The US government also unabashedly flexed its diplomatic muscles to get the US financial obligations to the United Nations reduced. The Bush administration also campaigned to undermine the ICC that was otherwise supported by most of the UN members. Many UN Member States therefore see the United States much more as a lone ranger than as a leader in UN affairs.

One should not ignore the appreciable contributions to the United Nations made by the United States. The United States built and led a coalition to conduct the first Gulf War, to deploy the peacekeeping mission in East Timor, to negotiate the end of North–South war in Sudan. The United States is well recognized for its efforts to build international will for intervention in Darfur, and in fact it was the only UN Member State that called the Darfur crisis a genocide. The United States is also the largest financier of the UN operations. The US Defense Department's 2006 Quadrennial Defense Review (QDR) highlighted its readiness to "increase its assistance to the UNPKO in areas of the Department's expertise, such as doctrine, training, strategic planning and management"[59] though this commitment has not been reflected in US priorities at the United Nations. The United States has not made improving genocide prevention capacities a focus of its reform efforts.

Time Magazine has observed the need for a more thoughtful and less parochial leadership.[60] Effective American leadership is welcomed. Speaking at Georgetown University to mark the tenth Anniversary of the fall of the Berlin Wall, President Clinton urged his successors to ensure America remained involved as the leader in world affairs.[61] A White House spokesman observed how "we have exerted leadership for the good and we need to continue in that tradition."[62] President Clinton admitted in February 1999 that Americans should "embrace the inexorable logic of globalization" and accept that "everything, from the strength of our economy to the safety of our cities, to the health of our people depended on events not only within our borders but half a world away."[63]

There is an increasing expectation of America to establish a proper functional stable international peaceful order as the only superpower in a unipolar world, because it benefits from a stable world order. The United States must be aware that a threat in any spot of the world is structurally linked to threats to its global interest and its allies. In a globalized society, and a world of interdependency where states' interests are structurally linked, US selective involvement in some wars is a strategic blunder. For example, wars and weak states are the breeding grounds of terrorism and associated piracy, of drug and human trafficking, and of money laundering that fund terrorism. It is therefore in the interests of America to be at the forefront of responding to wars in weak, or failed, states. The core reality is that the United States can offer effective global leadership if it does not prioritize realpolitik national interests over global collective interests when dealing with issues such as civilian protection. The truth is that in the contemporary globalized world, it is difficult to distinguish threats the United States faces from those facing other states, because all states face mounting common threats directly or indirectly. US objective leadership will lead to a new world order intolerant of blatant human rights violations. The central issue regarding America's leadership of the world is how it uses its powers to design appropriate strategies to achieve global goals. In the current world order, the United States cannot dominate other powers such as China and Russia. So the key strategy to effective American leadership is to collaborate with other powers to solve global problems. The United States must recognize that it does not have enough power to compel others to follow its lead. The United States must consider compromises in its formulation of common policies to ensure the cooperation and participation of other states. This multilateral problem-solving approach does not deny the US pursuit of its own essential interests because, as the only superpower, it has inherent advantages in bargaining to protect its interests and simultaneously solve common global problems. US power and global leadership can be fruitfully exercised to build a coalition of nations to collectively deal with common global threats.[64]

Effective leadership requires the interplay of entrepreneurial, intellectual, economic, and diplomatic leadership. Former US Secretary of Defense Robert Gates spoke of the need to enhance American soft power by "a dramatic increase in spending on the civilian instruments of national security—diplomacy, strategic communications, foreign assistance, civic action and economic reconstruction and development."[65] Entrepreneurial leadership requires the United States to focus on normative multilateral diplomatic skills and commitments to international cooperation.

Entrepreneurial leaders foster integrative bargaining through effective agenda setting, publicizing the importance of the issues at stake, and formulating innovative policies to overcome bargaining impediments to reach agreeable solutions.[66] The world faces a dilemma of human insecurity that need to be addressed through visionary ideas. Intellectual leadership promotes visions and ideas; farsightedness; and designs, plans, and strategies for collective action by the international community.[67] Leadership requires followership and the United States can elicit the support of other states by embracing normative or multilateral approaches in dealing with global issues. The United States must elevate the importance of the GA and reestablish the balance between the GA and the SC in dealing with civilian protection issues. It must endeavor to establish a new, dynamic, and more productive UN system, void of acrimonies and realpolitik divisions and stalemates.

Conclusion

Finally, there can be no more compelling and credible stance for the mission than to advocate for the most vulnerable. That role is the basis of the United Nations' moral core, and a powerful tool in winning over reluctant peacemakers, in speaking truth to the abusive, and in building credibility with both the local population and people worldwide who will give their support to those who speak for civilians caught in conflict. Such moral suasion can have a tangible result: this unbiased but firm stance will help deliver credence to the mission's authority and determination to use its impartiality against those who challenge its efforts. While this approach does not guarantee success, actions taken by peacekeeping missions to protect civilians will engender respect for the United Nations, and help deter violence against the most vulnerable in the future.

Appendix I

Table A2.1 UN peacekeeping missions in civil wars from 1956 to 2006

Country	Mission Observer= 1 Traditional = 2 Multi-dimensional = 3 "Robust" = 4	Mission type	Period
Angola	UNAVEM I	1	1988–1991
Angola	UNAVEM II	1	1991–1995
Angola	UNAVEM III	1	1995–1997
Angola	MONUA	1	1997–1999
Liberia	UNOMIL	1	1993–1997
Namibia	UNTAG	1	1989–1990
Sierra Leone	UNOMSIL	1	1998–1999
Dominican Republic	DOMREP	1	1965–1966
El Salvador	ONUSAL	1	1991–1995
Tajikistan	UNMOT	1	1994–2000
Lebanon	UNOGIL	1	December 1958
Yemen	UNYOM	1	1963–1964
Western Sahara	MINURSO	1	1991–
Georgia	UNOMIG	1	1993–2006
Central Africa Republic	MINURCA	2	1998–2000
Republic of Congo	ONUC	2	1960–1964
Rwanda	UNAMIR	2	1993–1996
Guatemala	MINUGUA	2	January 1997
Cambodia	UNAMIC	2	1991–1992
Cambodia	UNTAC	2	1992–1993
West New Guinea	UNSF	2	1962–1963
Sudan	UNMIS	2	2005–
Cyprus	UNFICYP	2	1964–

Table A2.1 (Continued)

Country	Mission Observer= 1 Traditional = 2 Multi-dimensional = 3 "Robust" = 4	Mission type	Period
Mozambique	ONUMOZ	3	1992–1994
Somalia	UNOSOM I	3	1992–1993
Haiti	UNSMIH	3	1996–1997
Haiti	UNTMIH	3	1997
Haiti	MIPONUH	3	1997–2000
East Timor	UNMIT	3	2006
Croatia	UNCRO	3	1995–1996
Croatia	UNTAES	3	1996–1998
Croatia	UNPSG	3	1998
Former Yugoslavian Republic	UNPROFOR	3	1992–1999
Macedonia	UNPREDEP	3	1995–1999
Bosnia and Herzegovina	UNMIBH	3	1996–2000
Haiti	MINUSTAH	3	2004–
Sierra Leone	UNAMSIL	4	1999–2005
Haiti	UNMIH	4	1993–1996
East Timor	UNTAET	4	1999–2002
Haiti	UNMISET	4	2002–2005
Somalia	UNOSOM II	4	1993–1995
Democratic Republic of Congo	MONUC	4	1999–
Liberia	UNMIL	4	2003–
Côte d'Ivoire	MINUCI	4	2003–2004
Côte d'Ivoire	UNOCI	4	2004–
Kosovo	UNMIK	4	1999–

Appendix 2

Table A3.1 Rules of engagement of traditional peacekeeping (UNAMIR, Rwanda)

The Security Council resolution 912 mandates UNAMIR

(a) to act as intermediary between the parties in attempt to secure their agreement to a ceasefire
(b) to assist in the resumption of humanitarian relief operations to the extent feasible
(c) to monitor and report on developments in Rwanda, including the safety and security of the civilians who sought refuge with UNAMIR.

Source: The Security Council resolution 912.

Table A3.2 Rules of engagement of "robust" peacekeeping (MONUC, DRC)

UN Security Council resolution 1291 mandates MONUC as follows:

MONUC may take the necessary action, in the areas of deployment of its infantry battalion and as it deems it within its capabilities to protect United Nations and collated JMC personnel, facilities, installations, and equipment, and ensure the security and freedom of movement of its personnel and protect civilians under imminent threat.

Source: Security Council resolution 1291.

Table A3.3 Rules of engagement enforcement action (Gulf War 1990)

The Security Council 678 (1990) (para 2)

Authorizes Member States cooperating with the Government of Kuwait, unless Iraq on or before January 15, 1991, fully implements, as set forth in paragraph 1 above, the above-mentioned resolutions, to use all necessary means to uphold and implement resolution 660 (1990) and all subsequent relevant resolutions and to restore international peace and security in the area.

Source: Security Council resolution 678.

Appendix 3

Table A3.4 Characteristics of UN mission types

Mission type	Rules of engagement	Weapon type	Force strength	Major power participation
Observer	No use of force	No arms	Small	No
Traditional	Use force only in self defense	Small arms (e.g., rifles)	Larger than observer	No
Multidimensional	Use force only in self defense	Small arms (e.g., rifles)	Larger than traditional mission	No/Yes
Robust	Use force for self defense and to protect civilians	Heavy arms (e.g., tanks)	Larger than traditional, observer and multidimensional mission	Yes
Enforcement	Maximum use of force to enforce peace	Heavier weapons (e.g., fighters aircrafts)	Larger than all missions	Yes

Notes

Chapter 1

1. Kofi Nsia-Pepra. 2011. "Robust Peacekeeping? Panacea for Human Rights Violations." *Peace and Conflict Studies* 18(2): 264.
2. Brahimi Report, 2000, United Nations, Report of the Panel on United Nations Peace Operations A/55/305.
3. Thakur Ramesh. 1994. "From Peacekeeping to Peace Enforcement: UN Operations in Somalia." *The Journal of Modern African Studies* 32(3): 387–410.
4. Fortna Virginia Page. 2004. "Does Peacekeeping Keep Peace?—International Intervention and the Duration of Peace after Civil War." *International Studies Quarterly* 48(2): 269–292.
5. Bercovitch Jacob. 1997. "Mediation in International Conflict: An Overview of Theory, a Review of Practicce." In *Peacekeeping in International Conflict: Methods and Techniques*, ed. I. W. Zartman and J. L. Rasmussen, Washington, D.C. United States: Institute of Peace; See also Brockner Joel. 1982. "Factors Affecting Entrapment in Escalating Conflicts." *Journal in Research in Personality* 6: 247–266.
6. Security Council Reso. 1265 among others expressed the Council's willingness to consider how peacemaking mandates might better address the negative impact of armed conflict on civilians—S/RES/1265 September 17 [14] [11]. Security Council 1296 April, On Protection of Civilians in Armed Conflict, affirmed the previous resolution addressing the need to better adequately provide resources to enhance peacekeepers' capacity to protect civilians under imminent threat-SC Res. 1296, April 2000.
7. See Security Council Documents, S/RES/2000, August 4, 2000, paragraph 3. http://www.un.org/Docs/scres/2000/2000sc1313.htm. Accessed: July 30, 2007.
8. Since Sierra Leone, the Council has authorized most UN-led peacekeeping missions with mandates primarily under Chapter VII "to protect civilians under imminent threat of physical violence" or a close variation thereof. Theses missions include Sierra Leone (UNAMSIL, 1999), the Democratic Republic of the Congo (MONUC, 1999), East Timor (1999), Liberia (UNMIL, 2003), Côte d'Ivoire (UNOCI), Burundi (ONUB), Haiti (MINUSTAH), Sudan (UNMIS, 2005), Darfur, Sudan (UNAMID, 2007/2008), Chad and Central African Republic (MINURCAT, 2007), and Lebanon (UNIFIL, 2007).

9. Lieberman Evans S. 2005. "Nested Analysis as a Mixed-Method Strategy for Comparative Research." *American Political Science Review* 99(3): 435–452.

Chapter 2

1. Preamble, United Nations Charter, http://www.un.org/en/documents/charter/.
2. Ibid.
3. Akhavan Payam. 1993. "Punishing War Crimes in the Former Yugoslavia: A Critical Juncture for the New World." *15 HRQ* 15(2): 264.
4. Brahimi Report. 2000. United Nations, *Report of the Panel on United Nations Peace Operations A/55/305.*
5. L. B. Curzon. 1998. *Dictionary of Law*, 5th Ed. London: Pitman.
6. Jack Donnelly. 2002. *Universal Human Rights*. Ithaca, New York USA: Cornell University Press, 57.
7. Henry Shue. 1980. *Basic Rights: Subsistence, Affluence, and U.S. Foreign Policy.* Princeton: Princeton University Press.
8. Weston Burns. 1995. "Human Rights." *Encyclopedia Britannica* 20 (Chicago): 713.
9. Ibid.
10. Stephen E. Lucas. 1989. "Justifying America: The Declaration of Independence as a Rhetorical Document." In *American Rhetoric: Context and Criticism*, ed. Thomas W. Benson, Carbondale, Illinois: Southern Illinois University Press, 85.
11. Declaration of the Rights of Man and of the Citizen [1789], http://www.britannica.com/bps/additionalcontent/8/116846/Declaration-of-the-Rights-of-Man-and-of-the-Citizen.
12. Burns, 174.
13. Edmund Burke and L. G. Mitchell. 2009. *Reflections on the Revolution in France.* NY, USA: Oxford University Press.
14. David Hume, David Fate Norton and Mary J. Norton. 2000. *Treatise of Human Nature.* NY, USA: Oxford University Press.
15. The Collected Works of Jeremy Bentham, ed. J. Bowring, London. 1843, 501.
16. Burke. 2006, *Reflections on the Revolution in France.*
17. The Collected Works of Jeremy Bentham, ed. J. Bowring.
18. Burns, 174.
19. Philip Alston, Ryan Goodman and Henry J. Steiner. 1996. *International Human Rights in Context: Law, Politics, Morals.* NY USA: Oxford University Press, 60.
20. Cornelius Flinterman and Jeroen Gutter. 2000. *The United Nations and Human Rights: Achievements and Challenges.* Paper-UNDP Publications.
21. Preamble, 1948 United Nations Charter, http://www.un.org/en/documents/charter/preamble.shtml.
22. See Article 1 UN Charter cited in J. Brownlie. 1999. *Basic Documents in International Law.* Oxford: Laredon.

23. Ibid.
24. R. Brad Roth. 1999. *Governmental Illegitimacy in International Law.* Oxford: Clarendon Press; New York: Oxford University Press.
25. P. Fitfoot v. 1992. "Functions and Powers and Inventions: UN Action in Respect of Human Rights and Humanitarian Intervention." In *To Loose the Bands of Wickedness*, ed. N. Rodley, United Kingdom: Brassey's (UK) Ltd.
26. Karel Vasak. 1977. "Human Rights: A Thirty-Year Struggle: The Sustained Efforts to Give Force of Law to the Universal Declaration of Human Rights." *UNESCO Courier, Paris: United Nations Educational, Scientific, and Cultural Organization* 30(11): 29.
27. Joshua S Goldstein and Jon C. Pevehouse. 2008. *International Relations*, 8th Ed. New York: Longman, 263–273.
28. Burns, 715.
29. Jack Donnelly. 2002. *Universal Human Rights.* Ithaca, New York USA: Cornell University Press, 58.
30. Burns, 716.
31. President Roosevelt's January 11, 1944, message to the Congress of the United States on the State of the Union. "The Economic Bill of Rights." Franklin D. Roosevelt American Heritage Center, http://www.fdrheritage.org/bill_of_rights.htm.
32. The European Social Charter is a Council of Europe treaty which guarantees social and economic human rights. It was adopted in 1961 and revised in 1996, Council of Europe, http://www.coe.int/T/DGHL/Monitoring/SocialCharter/.
33. Article 28 of the Universal Declaration of Human Rights.
34. 1972 Stockholm Declaration of the United Nations Conference on the Human Environment, http://www.unep.org/Documents.Multilingual/Default.asp?documentid=97&articleid=1503.
35. 1992 Rio Declaration on Environment and Development, http://www.unep.org/Documents.Multilingual/Default.asp?documentid=78&articleid=1163.
36. The Hungarian Parliamentary Commissioner for Future Generation http://jno.hu/en/.
37. Finland's Parliamentary Committee for the Future, http://web.eduskunta.fi/Resource.phx/parliament/committees/future.htx.
38. Organization for Security and Co-operation in Europe, http://www.osce.org/hcnm/.
39. Maurice Cranston. 1964. *What Are Human Rights*? New York. Basic Books, 40.
40. Ibid., 54.
41. Donnelley, 28.
42. Maurice Cranston. 1973. *What Are Human Rights?* New York: Taplinger Publishing, 65, 69.
43. Ibid., 66.
44. Cranston. 1964, 37.
45. Cranston. 1973, 66.
46. Ibid., 68.

47. Maurice Cranston. 1967. "Human Rights: Real and Supposed." In *Political Theory and the Rights of Man*, ed. D. D. Raphael, Bloomington: Indiana University Press, 43–51.
48. Burns, 716.
49. 1993 Vienna Convention on Human Rights, http://www.unhchr.ch/huridocda/huridoca.nsf/(symbol)/a.conf.157.23.en.
50. Preamble ICCPR, http://www2.ohchr.org/english/law/ccpr.htm.
51. The Declaration on the Right to Development adopted by the General Assembly of the UN on December 4, 1986, http://www.un.org/documents/ga/res/41/a41r128.htm.
52. Burns Weston. "Human Rights" from Encyclopedia Britannica. 15 ed176).
53. Burns, 717.
54. Donnelly, 71.
55. Preamble ICCPR, http://www2.ohchr.org/english/law/ccpr.htm.
56. 1948 Universal Declaration of Human Rights, http://www.un.org/en/documents/udhr/index.shtml.
57. 1968 Proclamation of Tehran, http://untreaty.un.org/cod/avl/ha/fatchr/fatchr.html.
58. In March 1993, 240 participants from 110 human rights and development organizations representing 26 countries from the Asia-Pacific region attended the Asia-Pacific NGO conference on human rights and issued a Joint Statement on the universality of human rights declaring, among others, that "the universal nature of human rights is beyond question; their promotion and protection are the duty of all states, regardless of their cultural, economic or political systems."
59. The 1993 Vienna Declaration provides in its paragraph 5 that "All human rights are universal, indivisible and interdependent and interrelated."
60. Many of these states that include non-Western states opposed to the universalists premise of human rights movement have ratified and incorporated the Universal Declaration of Human Rights into their Constitutions.
61. Jerome J. Shestack. 1998. "The Philosophical Foundations of Human Rights," *HRQ* 20(2): 201–234, 231.
62. Henry J. Steiner and Philip Alston. 1996. *International Human Rights In Context- Law, Politics and Morals*. Calderon: -Oxford University Press, 192.
63. Burns, in Steiner ed pa. 41.
64. Messer Ellen. 1997. "Pluralists Approaches to Human Rights." *Journal of Anthropological Research* 53(3): 293–317. Baxi also notes that "all human rights imagination is the estate of the west" (Upendra Baxi. 1999. "Voices of the Suffering, Fragmented Universality, and the Future of Human Rights." In *The Future of International Human Rights*, ed. Burns H. Weston and Stephen P. Marks, Ardsley, NY.: Transnational Publishers, 134).
65. Donnelley, 62.
66. Ibid.
67. Donnelley, 63.
68. Steiner, 193.

69. Makau Wa Mutua. 1995. "Why Redraw the Map of Africa: A Moral and Legal Inquiry." *Mich Int'l L 16*: 1113–1176.
70. Okere Obinna B. 1984. "The Protection of Human Rights in Africa and the African Charter: A Comparative Analysis with the European and American Systems." *Human Rights* Q.6(2): 141–159.
71. Mbiti John. 1990. *African Religions and Philosophy*. Portsmouth, New Hampshire, USA: Heinemann Educational Publishers, 141.
72. Kiwanuka Richard N. 1988. "The Meaning of 'People' in the African Charter on Human Rights and People's Rights." *82 Am. J. Int'l L.* 82(1) (January): 80–101.
73. Makau Wa Mutua. 1996. "The Ideology of Human Rights." *36 Virginia Journal of International Law* 589: 626–646.
74. Mutua, 643.
75. Adamantia Pollis and Peter Schwab. 1979. "Human Rights: A Western Construct with Limited Application." In *Human Rights: Cultural and Ideological Perspectives*, ed. Pollis and Schwab, New York: Praeger Publications, 1–17.
76. Joseph Nyerere. 1968. *Freedom and Socialism*. London: Oxford University Press, 1–12; see also Pollis and Schwab. "Human Rights: A Western Construct with Limited Application." 1–17.
77. 1993 Vienna World Conference on Human Rights, cited in Pattern Chris, East and West, Toronto, McClelland and Stewart Inc., 1998 page 150. See also Sen Amartya.1997. "Human Rights and Asian Values." *The New Republic*, July 14–21.
78. Xiaorong Li. 1999. "Asian Values and the Universality of Human Rights." *Business and Society Review* 102/103: 81–87, 81.
79. Bangkok Declaration 1967, http://www.aseansec.org/1212.htm.
80. Ibid.
81. Xiaorong, 82.
82. Yas Ghai. 1996. "Human Rights and Governance: The Asian Debate." , *International Human Rights in Context: Law, Politics, Morals*, ed. Philip Alston, Ryan Goodman and Henry J. Steiner, New York USA: Oxford University Press, 236–238. See also Susan Sim. 1996. "Human Rights: The East Asian Challenge." *The Straights Times*, Singapore February 18, quoted in Mutua et al. page 644 of Steiner.
83. Pollis and Schwab, 237.
84. Ghai, 644.
85. Article 1 UN Charter cited in Brownlie. *Basic Documents in International Law.*
86. D. J. Harris. 1998. *Cases and Materials on International Law*, 5th Ed. London, UK: Sweet and Maxwell, 6.
87. Roth Brad R. 2004. "The Enduring Significance of State Sovereignty." *Florida Law Review* 56(5) (December): 1023.
88. Ibid.
89. J. L. Briel.1925. "Matters of Domestic Jurisdiction." *British Yearbook of International Law* VI: 8.

90. The United States submitted a number of reservations before ratifying the ICCPR.
91. Article 1(2) of the UN Charter.
92. Paragraph 1,First Articles Common to ICCPR and ICESCR.
93. Kenneth N. Waltz. 1979. *Theory of International Politics*. Reading, MA: Addison-Wesley.
94. Article 39 UN Charter.
95. Roland Cox. 1994. *Power and Profits*. Lexington KY: University of Kentucky Press. See also Frieden A. Jeffrey. 1989. "The Economics of Intervention: American Overseas Investments and Relations with Underdeveloped Areas, 1890–1950." *Comparative Studies in Society and History* 31(1): 55–80.
96. Kaplan Robert D. as quoted in Walter A. McDougal. 1997. *Promised Crusader State: The American Encounter with the World Since 1776*. Boston: Houghton Mifflin, 207.
97. *Congressional Record-Senate*, 139(133), October 5, 1993, p. S13043 and 139(134), October 6, 1993, p. S13146.
98. United Nations Press Release SG/SM/6503 "Secretary General Says Proposals in his Report on Africa Require New Ways of Thinking, of Acting" (April 16, 1998, accessed January 30, 2007) Available at http://www.un.org/ecosocdev/genonfo/afrec/sgreport/pressrel.htm;internet.
99. Amnesty International Report. 2003.
100. Hultman Lisa. 2006. Civilians as Military Targets: Violent Bargaining Strategies by Governments and Rebels Presented at the Annual ISA Convention, San Diego, 22–25, March 13.
101. Ibid., 14.
102. Jordan J. Paust et al. 2001. *Human Rights Module: Crimes Against Humanity, Genocide, Other Crimes against Human Rights, and War Crimes*. Durham, North Carolina, USA: Corilina Academic Press, 130.
103. "Sexual violence" Available at: http://www.un.org/womenwatch/daw/public/cover.htm.
104. Cassasese Antonio. 1990. *Human Rights Achievements in a Changing World*. Philadelphia, Pennsylvania USA: Temple University Press, 190.
105. Ibid., 123.
106. Amnesty International. 1997. "Uganda-'Breaking God's Commands': The Destruction of Childhood by the Lord's Resistance Army." September 18.
107. William J. Durch, ed. 1993. *The Evolution of UN Pecacekeeping: Case Study and Comparative Analysis*. New York: St Martin Press.
108. See Cold War Canada, and Suez Crisis, retrieved on April 6, 2007 at file://E:/Suez Crisis.com.
109. See Diehl Paul. 1988. "Peacekeeping Operations and the Quest for Peace." *Political Science Quarterly* 103(Autumn): 485.
110. See Cold War Canada, Suez Crisis, at file://E:/Suez Crisis.com Accessed: June 4, 2007.
111. United Nations. 1990. *The Blue Helmets: A Review of United Nations Peacekeeping*, 2nd Ed. New York: The United Nations Department of Public Information, 5.

112. Thomas G. Weiss, David P. Forsythe and Roger A. Coate. 2001. *The United Nations and Changing World Politics*, 3rd Ed. Boulder, CO: Westview.
113. Marrack Goulding. 1993. "The Evolution of United Nations Peacekeeping." *International Affairs* 69(3): 451–464, 451.
114. Guehenno Jean Marie. 2002. "On the Challenges and Achievements of Reforming UN Peace Operations." *International Peacekeeping* 9(2): 69–80.
115. United Nations, www.un.org/Depts/dpko/dpko.
116. Indar J. Rihkye. 1984. *The Theory and Practice of International Peacekeeping*. London, UK: Hurst and Company, 1–2.
117. M. Goulding. 1993. "The Evolution of United Nations Peacekeeping." *International Affairs* 69(3): 452.
118. John Mackinlay and Jarat Chopra. 1992. "Second Generation Multinational Operations." *The Washington Quarterly* 15(3): 113–131.
119. Fortna Virgina Page. 2004. "Does Peacekeeping Keep Peace? International Intervention and the Duration of Peace After Civil War." *International Studies Quarterly* 48(2): 269–292; Brahimi Report. 2000.
120. Paul F. Diehla, Jennifer Reifschneidera and Paul R. Hensel. 1996. "United Nations Intervention and Recurring Conflict." *International Organization* 50(4): 683–700.
121. Roberts Adams. 1994. "The Crisis in Peacekeeping." *Survival* 36(3): 115.
122. Duane Bratt. 1997. "Explaining Peacekeeping Performance: The UN in Internal Conflicts." *International Peacekeeping* 4(3): 45–70, 66.
123. P. F. Diehl. 1993. *International Peacekeeping*. Baltimore, USA: The Johns Hopkins University Press.
124. See Boutros-Ghali. 1992. *An Agenda for Peace*. New York: United Nations Department of Public Information.
125. Claude Jr Inis L. 1956. *Swords into Plowshare: The Problem and Progress of International Organization*. New York: Random House.
126. Dominick Donald. 2003. "Neutral Is Not Impartial: The Confusing Legacy of Traditional Peace Operations Thinking." *Armed Forces and Society* 29: 415.
127. Thakur Ramesh. 1994. "From Peacekeeping to Peace Enforcement: UN Opeartion in Somalia." *The Journal of Modern Africa Studies* 32(3): 387–419, 399.
128. Stuart Griffin. 2001. "Peacekeeping, the United Nations, and the Future Role of Commonwealth." *Commonwealth and Comparative Politics* 39(3): 150–164.
129. James Burk. 2000. "What Justifies Peacekeeping?" *Peace Review* 12(3): 468.
130. See Brahimi Report. 2000.
131. Preamble, United Nations Charter.
132. Daniel Druckman, Stern Paul C., Diehl P., Fetherston Betts A., Johansen Robert, Durch William and Ratner Steven. 1997. "Evaluating Peacekeeping?" *Mershon International Studies Review* 41(1): 151–165.
133. Victoria Holt, Glyn Taylor and Max Kelly. 2009. "UN Peacekeeping Operations Successes, Setbacks and Remaining Challenges" Independent study

jointly commissioned by the Department of Peacekeeping Operations and the Office for the Coordination of Humanitarian Affairs, New York: United Nations, 22.

134. Ibid., 24; see Preamble of the UN Charter.
135. Siobhan Willls. 2009. *Protecting Civilians- The Obligations of Peacekeepers*. USA: Oxford University Press, 1 edition, 9.
136. Annual Report of the Secretary General on the Work of the Organization, June 16, 1960–June 1961, 16th Session, UN Doc A/800, 27; A. Sitowski. 2006. *UN Peacekeeping Myth and Reality.* London and Westport: Praeger Security International.
137. See, for example, Rwanda: The Preventable Genocide (IPEP, 2000); Report of the Secretary-General Pursuant to General Assembly Resolution 53/35: The Fall of Srebrenica A/54/549 of November 15, 1999. Report of the Independent Inquiry into the actions of the United Nations during the 1994 genocide in Rwanda, S/1999/1257 of December 16, 1999.
138. Report of the Secretary-General Pursuant to General Assembly Resolution 53/35: The Fall of Srebrenica A/54/549 of November 15, 1999.
139. Ibid., 503.
140. Report of the Panel on the United Nations Operation A/55/305-S/2000/809 August 2000, Hereafter the Brahimi Report. 2000, 2.
141. Ibid., 620.
142. Daniel Druckman and Paul C. Stern. 1997. "Evaluating Peacekeeping?" *Mershon International Studies Review* 41(1): 151–165.
143. Thomas G. Weiss, ed. 1995a. "The United Nations and Civil Wars at the Dawn of the Twenty-First Century." In *The United nations and Civil Wars*, ed. Thomas G. Weiss, Boulder, CO: Lynne Reinner.
144. Paul F. Diehl. 1994. *International Peacekeeping*. Baltimore, MD: Johns Hopkins University Press.
145. Johansen Robert C. 1994. "UN Peacekeeping: How Should Measure Success?" *Mershon International Studies Review* 38: 307–310.
146. Daniel et al. "Evaluating Peacekeeping?" 163.
147. A. B. Fetrerston. 1994. *Towards a Theory of United Nations Peacekeeping.* New York: St Martins's Press.
148. G. L. Sherry. 1986. "The United Nations, International Conflict, and American Security." *Political Science Quarterly* 101(5): 753–771.
149. Thakur Ramesh. 1994. "From Peacekeeping to Peace Enforcement: UN Operation in Somalia." *The Journal of Modern Africa Studies* 32(3): 387–419, 399.
150. Allan James. 1990. "International Peacekeeping: The Disputants' View." *Political Studies* 38(2): 215–230.
151. Bruce Russett and J. S. Sutterlin. 1991. "The UN in a New World Order." *Foreign Affairs* 70(2): 69–83.
152. John Hillen. 2000. *Blue Helmets: The Strategy of UN Military Operations*, 2nd Ed. Washington, D.C.: Brassey's.
153. UN forces stood by as Serbs massacred Bosnian Muslims. (about 7,000 victims in Srebrenica); see Stephen Edwards. 2003. "The Failure of the

Security Council a Blow to Chrétien Government." *National Post*, Wednesday, March 16, A9; See also Duane Bratt. 1996. "Assessing the Success of UN peacekeeping Operations." *International Peacekeeping* 3(4): 66–83.

154. Diehl. *International Peacekeeping*.
155. William J. Durch, ed. 1993. *The Evolution of UN Peacekeeping: Case Studies and Comparative Analysis*. Washington, D.C.: The Henry L. Stimson Center, 12.
156. Diehl. *International Peacekeeping*.
157. Malaquias Assis. 1996. "The UN in Mozambique and Angola: Lessons Learned." *International Peacekeeping* 3: 88.
158. Chantal de Jonge Qudraat. 1996. "The United Nations and Internal Conflict." In *The International Dimensions of Internal Conflict*, ed. E. Michael, Brown Cambridge, MA: MIT Press.
159. Edward. N. Luttwak. 1999. "Given War a Chance." *Foreign Affairs* 78(4): 36–44; see also Shashi Thardoor. 1995. "Should United Nations Peacekeeping Go 'Back to Basics'?" *Survival* 37(2): 52–64.
160. Donald C. F. Daniel and Bradd C. Hayes. 1996. "Securing Observance of UN Mandates through the Employment of Military Force." *International Peacekeeping* 3(4): 105–125.
161. Thomas R. Mockaitis. 1999. *Peace Operations and Intrastate Conflict: The Sward or the Olive Branch?* Westport, CT: Praeger.
162. John Hillen. 2000. *Blue Helmets: The Strategy of UN Military Operations*. 2nd Ed. Washington, D.C.: Brassey's.
163. Ernst B. Hass. 1986. *Why We Still Need the United Nations: The Collective Management of International Conflict, 1945–1985*. Policy Papers in International Affairs, No. 26. Berkeley, Calif.: Institute of *International* Studies.
164. Michael W. Doyle and Nicholas Sambanis. 2000. "International Peacebuilding: A Theoretical and Quantitative Analysis." *American Political Science Review* 94(4): 779–801.
165. Hartzell Caroline, Mathew Hoddie and Donald Rothcild. 2001. "Stabilizing the Peace after Civil War." *International Organizations* 55: 183–208.
166. A. Dubey. 2002. "Domestic Institutions and the Duration of Civil Wars Settlement." Paper presented at the Annual Meeting ISA New Orleans.
167. George W. Downs and Stephen John Stedman. 2002. "Evaluation Issues in Peace Implementation." In *Ending Civil Wars: The Implementation of Peace Agreements*, ed. Stephen John Stedman, Donald Rothchild and Elizabeth M.Cousens, Boulder, CO: Lynne Rienner.
168. See Doyle and Sambanis. 2000. "International Peacebuilding: A Theoretical and Quantitative Analysis." 779–801; Virginia Page Fortna. 2003. "Inside and Out: Peacekeeping and the Duration of Peace after Civil and Interstate Wars." *International Studies Review* 5(4): 97–114.
169. See International Relations and Security Network Report 2005.
170. In Sudan, by way of contrast, the local forces and militia combined had more than 50,000 troops, outnumbering the 6,200 peacekeeping forces.

Chapter 3

1. Preamble, United Nations Charter, http://www.un.org/en/documents/charter/.
2. Preamble, United Nations Charter.
3. Lacey Marc. 2005. "UN Forces Using Tougher Tactics to Secure Peace." *New York Times*, May 23.
4. Patrick Regan. 1996. "Conditions for Successful Third Party Interventions." *Journal of Conflict Resolution* 40(1): 336–359.
5. Preamble, United Nations Charter.
6. Wills Siobhan. 2009. *Protecting Civilians: The Obligations of Peacekeepers.* New York, USA: Oxford University Press, 9.
7. Annual Report of the Sec-General on the Work of the Organization, June 16, 1960–June 1961, 16th session, UN Doc a/800, 27; A. Sitowski. UN Peacekeeping Myth and Reality (Prager Security International London and Westport 2006, also footnotes 44 page 11 of Willis).
8. Ibid.
9. Berdal Mats. 2003. "Ten Years of International Peacekeeping." *International Peacekeeping* 10(4): 10.
10. Ramesh Thakur and Carlyle Thayer, eds. 1995. *A Crisis of Expectations: UN Peacekeeping in the 1990s*. Boulder, CO: Westview Press, 317.
11. Preamble, United Nations Charter.
12. Stian Kjeksrud. 2009. "Matching Robust Ambitions with Robust Action in Operations- Towards a Conceptual Overstretch?" *Norwegian Defence Research Establishment (FFI)* April 20 (A Report), 12.
13. B. Boutrous-Ghali. 1992. *Agenda for Peace.* New York: United Nations.
14. Ibid., para 9.
15. Ibid., para 42, 43.
16. Comprehensive Review of the Whole Question of Peacekeeping Operations in all their Aspects; Report of the Secretary General on Implementation of the Recommendations of the Special Committee on Peacekeeping Operations and the Panel on United Nations Operations A/55/977 June 1, 2001, 65; See Willis footnote 213, 38.
17. UNDPO. 1996. Lessons Learned Unit, Department of UN Peacekeeping Operations, Multidisciplinary Peacekeeping: Lessons from Recent Experience, December; see Willis footnote 214, 39.
18. UNGA. 1999. Report of the Secretary General pursuant to General Assembly Resolution 53/55: The Fall of Srebrenica; (November 15, 1999); Report of the Independent Inquiry into the Actions of the United Nations during the 1994 Genocide New York United Nations December 15, 1999.
19. UNGA. 1999. Report of the Independent Inquiry into the Actions of the United Nations during the 1994 Genocide New York United Nations December 15, 1999, para 43.
20. UNGA. 1999. Report of the Secretary General pursuant to General Assembly Resolution 53/55: The Fall of Srebrenica; (November 15, 1999).
21. Ibid., 505.

22. Annan. 1999a. "Walking in the International Tightrope." *The New York Times* (September): 19; Annan. 1999b. "Two Concepts of Sovereignty." *The Economist* (September 18): 49–50.
23. Annan. 1999a.
24. Ibid.
25. R. Dallaire. 2003. *Shaking Hands with the Devil: The Failure of Humanity in Rwanda.* Canada: Random House.
26. Annan Two Concepts of Sovereignty.
27. UNGA. 1999. *Report of the Secretary-General to the Security Council on the Protection of Civilians in Armed Conflict* S/1999/957, September 8, 1999, para. 57.
28. UNGA. 2000. *We the Peoples': The Role of the United Nations in the 21st Century*, Millennium Report of the Secretary-General, GA/9704 A/54/2000 of March 27, 2000.
29. Kofi Annan.(April 13, 1998). "Report of the Secretary-General on the situation in Africa, Kofi Annan. 1999. Report of the Secretary General on the Protection of Civilians in Armed Conflict (September 8, 1999).
30. UNSC Res. 1265, 1999, Adopted by the Security Council at its 4046th meeting, on September 17, 1999 S/RES/1265 (1999).
31. Ibid.
32. UNSC, Resolution 1296 (2000) S/RES/1296 (2000) Adopted by the Security Council at its 4130th meeting, on April 19, 2000.
33. United Nations Security Council resolution 1270, (1999), S/RES/1270 (1999) October 22, 1999.
34. United Nations. Report of the Panel on United Nations Peace Operations, UN Doc. A/55/305-S/2000/809, August 21, 2000 (Hereafter Brahimi Report) paras 48–50, 55, 66d.
35. Ibid.
36. Ibid.
37. The Responsibility to Protect, 2001, Report of the International Commission on Intervention and State Sovereignty, ICISS.
38. Ibid.
39. He told a 1998 audience at Ditchley Park, Conference Center, England.
40. United Nations. 2004. *High Level Panel on Threats Challenges and Change: A More Secure World: Our Responsibility.* New York: United Nations.
41. Ibid., para. 203.
42. UN DPKO and DFS, 2008, *United Nations Peacekeeping Operations, Guidelines and Principles for Peacekeeping Operations*, New York: United Nations.
43. Ibid.
44. See UNSCR Resolution 1289 (2000) S/RES/1289 (2000), February 7, 2000, para 10.
45. Yap Ong Heng, Military Advisor, Singapore Mission to the United Nations, Statement at the Special Political and Decolonization Committee, on Comprehensive Review of the Whole Question of Peacekeeping Operations in all their Aspects, October 20, 2003.
46. Brahimi Report. 2000. Supra no. 7.

47. Secretary-General Kofi Annan in a remark delivered on UN Peacekeeping Operations to the Security Council 8/22/2005 Press Release (SG/SM/9311 SC/8096 PKO/107).
48. United Nations. 2005. World Summit Outcome, para. 92 obtained at http://www.un-ngls.org/un-summit-FINAL-DOC.pdf (August 22, 2007).
49. Ruggie John Gerard. 1993. "Wandering in the Void: Charting the UN's New Strategic Role." 72 *Foreign Aff.*, November–December: 26, 29.
50. See United States Department of State, *Administration Policy on Reforming Multilateral Peace Operations,* April 1994, 33 ILM 705, 706 at 809–10 [hereinafter US DEPT. OF STATE]
51. Donald Wittman. 1979. "How a War Ends: A Rational Model Approach." *Journal of Conflict Resolution* 23(4): 743–763.
52. L. Woodhouse. 1999. "The Gentle Hand of Peace? British Peacekeeping and Conflict Resolution in Complex Political Emergencies." *International Peacekeeping* Summer. 6(2): 31.
53. 58th General Assembly Meeting, Fifth Committee 19th Meeting, Press release GA/AB/359 2005.
54. Brahimi Report. 2000; United Nations. 2000.
55. United Nations. 2005. *Report of the Special Committee Peacekeeping and Its Working Group at the 2005 Substantive Session,* A/59/19, March 1, para. 30–46.
56. International Crisis Group. "UN Darfur Needs Robust Protection Force." May 2006.
57. Dale Stephens. 2005. "The Lawful Use of Force by Peacekeeping Forces: The Tactical Imperative." *International Peacekeeping* 12(2), Summer: 157–172, 157.
58. See United States Department of State, *Administration Policy on Reforming Multilateral Peace Operations,* April 1994, 33 ILM 705, 706 at 809–10 [hereinafter US DEPT. OF STATE]. See also Eric P. Schwartz. Consultant, Council on Foreign Affairs *UN Peacekeeping Reform: Seeking Greater Accountability and Integrity.* Testimony before the house of international relations, subcommittee on Africa, global human rights, and international organizations May 18, 2005 obtained on January 26, 2007, at, http://.cfr.org/publications/8113/un_peacekeeping_reform.html.
59. Special Committee on Political and Decolonization, 58th General Assembly, Fourth Committee tenth meeting, October 17, 2003.
60. Gen Dallaire Romeo. 1995. "Towards a Rapid Reaction Capability for the United Nations." *Canadian Department of Foreign Affairs*, Ottawa, September.
61. Roth Kenneth. 2006. "UN: Darfur Needs Protection Force." *Human Rights Watch* (May 25), 18(9).
62. UNAMIR's Force Commander, Maj.-Gen. Romeo Dallaire Supra No. 50, 5.
63. Sakda Sangsnit. 2005. At the 58th General Assembly meeting, fifth committee, 19th meeting Press release GA/AB/3592.
64. Ibid.
65. Ibid.

66. Guehenno. The Under-Secretary General for Peacekeeping in a press release to the Special Committee on Peacekeeping Operations 180th and 181st meetings of the General Assembly (GA/PK/183 in 2005).
67. The Brahimi Report makes clear that, noting that post–Cold War missions tended to deploy.

> where conflict had not resulted in victory for any side, where a military stalemate or international pressure or both had brought fighting to halt but a least some of the parties to the conflict were not seriously committed to ending the confrontation. United Nations operations thus did not deploy into post-conflict situations but tried to create it.

See United Nations, Report of the Panel of Peace Operations, 2000, viii.
68. Brahimi Report. 2000. Supra no. 7.
69. Wole Soyinka. 2000. "A Fate Worse than Imperialism." *The Guardian*, Saturday review, June 3.
70. See International Commission Report 2001 titled "Responsibility to Protect" (R2P) accessed at http://www.iciss.ca/pdf/Commission-Report.pdf (August 22, 2007).
71. He told a 1998 Audience at Ditchley Park, Conference Center England.
72. Brahimi Report. 2000 .
73. Preamble UN Charter, Victoria Holt and Glyn Taylor with Max Kelly. 2009. *UN Peacekeeping Operations Successes, Setbacks and Remaining Challenges*, Independent study jointly commissioned by the Department of Peacekeeping Operations and the Office for the Coordination of Humanitarian Affairs, United Nations, New York, 24.
74. Holts and Taylor, 23.
75. See Fifth report of the Secretary General on the United Nations Mission in Sierra Leone, S/2000/751 July 31, 2000, paragraph 12, http://www.un.org/Docs/sc/reports/2001//751e.pdf. Accessed: July 30, 2007.
76. See Security Council Documents, S/RES/2000, August 4, 2000, paragraph 3, http://www.un.org/Docs/scres/2000/2000sc1313.htm. Accessed: July 30, 2007.
77. Ruggie John Gerrad. 1993. "Wandering in the Void: Chartering the UN's New Strategic Role." *Foreign Affairs* 72(4): 26.
78. See Article 42 of the UN Charter.
79. UN Security Council resolution 1291 (2000) of February 24, 2000 [S/RES/1291 (2000)].
80. Ian Johnstone. 2006. "Dilemma of Robust Peace Operations." In *Robust Peacekeeping: The Politics of Force*. New York, US: Center on International Cooperation New York University, 66.
81. See Article 2(4) of the UN Charter.
82. See Article 39 of the UN Charter.
83. Parker Nealin J. 2009. *Robust Peacekeeping: The Politics of Force*. New York, US: Center on International Cooperation New York University, 2–6.

84. DPKO and Department of Field Support's Guidelines and Principles for Peacekeeping Operations, (*Capstone Doctrine*).
85. Johnstone, 66. Brahimi Lakhdar and Ahmed Salman. 2008. *Pursuit of Peace. The Seven Deadly Sins of Mediation in Robust Peacekeeping: The Politics of Force.* New York, US: Center on International Cooperation New York University.
86. C. D. Coning, C. Aoiand and R. Thakur. 2007. *Unintended Consequences of Peacekeeping Operations.* New York, US: United Nations University Press.
87. K. Allen. 2007. "Congo Peacekeepers Stuck in Middle." *BBC News.* http://news.bbc.co.uk/2/hi/africa/7064497.stm
88. Guehenno Jean Marie. 2005. The Under-Secretary General for Peacekeeping Press Release to the Special Committee on Peacekeeping Operations 180th & 181th meetings of the General Assembly (GA/PK/183 in 2005).
89. Gene Sharp. 1985. *Making Europe Unconquerable: The Potential Civilian-Based Deterrence and Defence.* London: Taylor and Francis.
90. Dale Stephens. 2005. "The Lawful Use of Force by Peacekeeping Forces: The Tactical Imperative." *International Peacekeeping* 12(2) Summer: 157.
91. John Ruggie. 1993. "Wandering in the Void: Charting the UN's New Strategic Role." *Foreign Affairs* 72(5): 26–31.
92. Gowan Richard and Benjamin Tortolani 2008, "Robust Peacekeeping and its Limitations" In *Robust Peacekeeping: The Politics of Force*, New York, US: Center on International Cooperation, New York University, 49.
93. Siekmann Robert. 1991. *National Contingents in United Nations Peacekeeping Forces.* Dordrecht: Nijhoff, 132.
94. Stephens, 159.
95. Rowe Peter. 1994. "The United Nations Rules of Engagement and the British Soldier in Bosnia." *International and Comparative Law Quaterly* 43: 955.
96. Bruce D. Jones. 2001. *The Best Laid Plans . . . Peace-Making in Rwanda and the Implication of Failure.* New York: Sage, 23.
97. General Sir Michael Jackson. 2008. *The Use of Force-General Reflections,* IFS Seminar, Oslo: Institute of Defense Studies, February 5.
98. William J. Durch. 1993. *The Evolution of UN Peacekeeping: Case Study and Comparative Analysis.* New York: St. Martins Press, 46.
99. Kjeksrud, 32.
100. Ibid., 31.
101. Kofi Annan. 2000. *We the Peoples: The Role of the United Nations in the 21st Century.* New York: United Nations.
102. Roméo Dallaire and Samantha Power. 2004. *Shake Hands with the Devil: The Failure of Humanity in Rwanda.* Cambridge, Massachusetts, USA: Da Capo Press, 99–100.
103. UN Report on Integrated Mission 2006 accessed at, http://64.233.167.104/search?q=cache:YYgZsbeknxYJ:ochaonline.un.org/OchaLinkClick.aspx%3Flink%3Docha%26DocId%3D1003352+report+on+integrated+missions&hl=en&ct=clnk&cd=1&gl=us (August 22, 2007).
104. Ibid.
105. Kjeksrud, 7.

106. Ibid.
107. W. J. Durch, ed. 2006. *Twenty-First Century Peace Operations*. Washington, DC, USA: United Institute of Peace, 582.
108. Wittman Donald. 1979. "How a War Ends: A Rational Model Approach." *Journal of Conflict Resolution*; See also David Mason and Patrick J. Fett. 1996. "How Civil War Ends: A Rational Choice Approach." *Journal of Conflict Resolution* 40(4) (December): 546–568.
109. Bruce Bueno de Mesquita and David Lalman. 1992. *War and Reason: Domestic and International Imperatives*. New Haven: Yale University Press.
110. Douglas Lemke and Jacek Kugler. 1996. "The Evolution of the Power Transition perspective." In *Parityand war. Evaluations and Extensions of the War Ledger*, ed. Jacek Kugler and Douglas Lemke, Ann Arbor: Michigan University Press, 3–33.
111. Ibid., 4.
112. James D. Fearon. 1995. "Rationalist Explanations for War." *International Organization* 49(3): 379–414.
113. William Reed. 2003. "Information, Power, and War." *American Political Science Review* 97(4): 633–641, 637.
114. Goldstein Joshua S. 2005. *International Relations*. New York: Pearson-Longman, 107.
115. Kenneth Waltz. 1959. *Man, the State, and War: A Theoretical Analysis*. New York: Columbia University Press.
116. Ibid., 232.
117. Bruce Bueno de Mesquita and David Lalman. 1992. *War and Reason. Domestic and International Imperatives*. New Haven, CT: Yale University Press. See also John R. Oneal, Frances H. Oneal, Zeev Maoz and Bruce M. Russett. 1996. "The Liberal Peace: Interdependence, Democracy, and International Conflict, 1950–85." *Journal of Peace Research* 33(1): 11–28; Havard Hegre 2004. "Size Asymmetry, Trade, and Militarized Conflict." *Journal of Conflict Resolution* 48(3): 403–429.
118. Regan 1996.
119. Shamar Boas Elron and Ben-Ari Eyal. 1999. "Why Don't They Fight Each Other? Cultural Diversity and Operational Unity in Multinational Forces." *Armed Forces and Society* 26(1): 73–97; T. Duffey. 2000. "Cultural Issues in Contemporary Peacekeeping." *International Peacekeeping* 7(1): 142–168.
120. P. F. Diehl. 1994. *International Peacekeeping,* Baltimore. USA: The Johns Hopkins University Press, 124.
121. Ibid., 128–129.
122. D. Lake and D. Rothchild. 1996. "Containing Fear: The Origins and Management of Conflict." *International Security*. 21(2): 41–75.
123. Patrick Regan. 2000. *Civil Wars and Foreign Powers*. Ann Arbor, Michigan, USA: University of Michigan Press; George Downs and Stephen John Stedman. 2002. "Evaluating Issues in Peace Implementation." In *Ending Civil Wars: The Implementation of Peace Agreements*, ed. Stephen John Stedman, Donald Rothchild, and Elizabeth M. Cousens, Boulder, Colorado: Lynne Rienner Publishers.

124. Havard Hegre. 2004. "Size Asymmetry, Trade, and Militarized Conflict." *Journal of Conflict Resolution* 48(3): 403–429.
125. Christian Davenport. 1999. "Human Rights and the Democratic Proposition." *Journal of Conflict Resolution* 43(1): 92–116.
126. R. J. Rummel. 1995. "Democracy, Power, Genocide, and Mass Murder." *Journal of Conflict Resolution* 39(1): 3–26.

Chapter 4

1. Indar Jit Rikhye. 1984. *The Theory and Practice of Peacekeeping*. London: C. Hurst & Co; William J. Durch. 1993. *The Evolution of UN Peacekeeping: Case Study and Comparative Analysis*. New York: St. Martins Press; P. F. Diehl. 1994. *International Peacekeeping*. Baltimore, USA: The Johns Hopkins University Press; Bratt Duane. 1997. "Explaining Peacekeeping Performance: The UN in Internal Conflicts." *International Peacekeeping* 4(3): 45–70; Justin Conlon. 2004. "Sovereignty vs. Human Rights or Sovereignty and Human Rights?" *Race and Class* 46(1): 76–100; Trevor Findlay. 2002. *The Use of Force in UN Operations*. Oxford, UK: Oxford University Press.
2. Christopher H. Achen and Duncal Snidal. 1989. "Rational Deterrence Theory and Comparative Case Studies." *World Politics* 41(2): 143–169; Gary King, Robert O Keohane, and Sidney Verba. 1994. *Designing Social Inquiry: Scientific Inference in Qualitative Research.* Princeton, New Jersey USA: Princeton University Press.
3. Paul Collier, Anke Hoffler, and Mans Soderbom. 2004. "On The Duration of Civil War." *Journal of Peace Research* 41(3): 253–273.
4. B. Hass Ernst. 1986. *Why We Still Need the United Nations: The Collective Management of International Conflict, 1945–1985*, Policy Papers in International Affairs. Berkley, California: Institute of International Studies, University of California; Andrew Gelman and Gary King. 1994. "A Unified Method of Evaluating Electoral Systems and Redistricting Plans." *American Journal of Political Science* 38(2): 514–554; Diehl Paul, Jennifer Reifschneder, and Paul R. L. Hensen. 1996. "United Nations Intervention and Recurring Conflict." *International Organization* 50(4): 683–700; Virginia Page Fortna. 2003. "Inside and Out: Peacekeeping and Duration of Peace After Civil Wars and Interstate Wars." *International Studies Review* 5(4): 97–114.
5. Christopher H. Achen and Duncal Snidal. 1989. "Rational Deterrence Theory and Comparative Case Studies." *World Politics* 41(2): 143–169; Evans Lieberman S. 2005. "Nested Analysis as a Mixed-Method Strategy for Comparative Research." *American Political Science Review* 99(3): 435–452.
6. Lieberman 2005, 435.
7. Peter Wallensteen and Margareta Sollenberg. 2001. "Armed Conflict 1989–2000." *Journal of Peace Research* 38(5): 629–644.
8. David J. Singer and Melvin Small. 1994. *Correlates of War Project: International and Civil War Data, 1816–1992,* ICPSR Study # 9905. Ann Arbor, Michigan, USA; Patrick Regan. 2000. *Civil Wars and Foreign Powers*. Ann Arbor, Michigan, USA: University of Michigan Press.

9. UCDP One-Sided Violence Dataset v.1.2 1989–2005 at, http://www.pcr.uu.se/research/UCDP/our_data1.htm. Accessed: December 4, 2007.
10. Ibid.
11. United States Department of State Annual Country Reports on Human Rights Practices, http://www.state.gov/r/pa/ei/a2z/26394.htm. Accessed: December 4, 2007.
12. Human Rights Watch Country Report, http://www.hrw.org/countries.html. Accessed: December 4, 2007.
13. Regan 2000; Political Instability Task Force Data (State Failure) 2007.
14. Scott J. Long. 1997. *Regression Model for Categorical and Limited Dependent Variables*. Thousand Oaks, California, USA: Sage Publishers.
15. Gary King. 1989. "Event Count Models for International Relations: Generalizations and Applications." *International Studies Quarterly* 33: 123–147.
16. Art E. James, Gary King and Cutis S. Signorino 2000. "Aggregation among Binary, Count, and Duration Models: Estimating the Same Quantities from Different Levels of Data." *Political Analysis* 9(1): 1–24.
17. King, 1989, 29; Long, 1997, 230–236.
18. Eck Kristine and Joakim Kreutz. 2006. Uppsala Conflict Data Program, One-Sided Violence Dataset 2006 version, http://www.pcr.uu.se/research/UCDP/data_and_publications/datasets.htm Accessed: June 27, 2007; Long, 1997, 242; King, 1989, 129; Volker Krause, Susumu Suzuki, and Richard Witmer. 2006. "US Domestic Politics and International Cooperation, 1953–1998." *International Politics* 43: 620–643.

Chapter 5

1. The variable arms type dropped due to perfect collinearity with rules of engagement. It is probably because both variables are coded in dummy format.
2. Peace agreement dropped due to collinearity.
3. Coefficients of negative binomial regression model give us odd values. In order to understand the actual impact of each coefficient, we need to compute the substantive value of each coefficient of the significant variable.
4. Model reduces errors in the prediction of change in civilian killings by 32.26 percent.
5. J. L. Snyder. 2000. *From Voting to Violence: Democratization and Nationalist Conflict*. New York, USA: W. W. Norton and Company, 320.

Chapter 6

1. George Alexander L. 1979. "Case Studies and Theory Development: The Method of Structured, Focused Comparison." In *Diplomacy New Approaches in History, Theory, and Policy*, ed. Lauren Paul Gordon, New York: The Free Press.
2. Martin W. James III. 2011. *A Political History of the Civil War in Angola 1974–1990*. New Jersey, USA: Transaction Publishers.

3. Jacqueline Audrey Kalley. 1999. *Southern African Political History: A Chronology of Key Political Events from Independence to Mid-1997*. Westport, CT and London: Greenwood, 36, See also Abiodun Alao. 1994. *Brothers at War: Dissidence and Rebellion in Southern Africa*. London: British Academic Press, XIX–XXI.
4. Inge Tvedten. 1997. *Angola: Struggle for Peace and Reconstruction*. Boulder, Colorado, USA: Westview Press, 38–39.
5. Karel C. Wellens. 1990. *Resolutions and Statements of the United Nations Security Council (1946–1989): A Thematic Guide*. Dordrecht, The Netherlands: Martins Nijhoff Publishers, 235–236.
6. In 1981 Chester Crocker, US assistant secretary of state for African affairs for newly elected president Ronald Reagan had developed a linkage policy, tying apartheid South Africa's agreement to retreat from Angola and to relinquish control of Namibia in line with United Nations Security Council resolution 435 to Cuba's troop withdrawal from Angola. On September 10, 1986, Fidel Castro basically accepted Crocker's proposal. The negotiations were finalized in New York with Angola, Cuba, and South Africa signing the accord on December 22, 1988. It supplied for the retreat of South African forces from Angola, which had already taken place by August 30, the withdrawal of South Africa from Namibia and Namibia's independence and the withdrawal of Cuban forces from Angola within 30 months. The New York Accords granted independence to Namibia and ended the direct involvement of foreign troops in the Angolan Civil War. See Piero Gleijeses. 1988. *Conflicting Missions: Havana, Washington, and Africa, 1959–1976*. Chapel Hill, North Carolina: The University of North Carolina Press quoting: Crocker to Secretary of State, June 26, 1988.
7. National Society for Human Rights, *Ending the Angolan Conflict*, Windhoek, Namibia, July 3, 2000 (opposition parties, massacres); John Matthew, Letters, *The Times*, UK, November 6, 1992 (election observer); NSHR, Press Releases, September 12, 2000, May 16, 2001 (MPLA atrocities).
8. Ibid.
9. Donald S. Rothchild. 1997. *Managing Ethnic Conflict in Africa: Pressures and Incentives for Cooperation*. Washington, DC: The Brookings Institution, 251.
10. Ibid., 137–138.
11. Martin W. James III. 2011. *A Political History of the Civil War in Angola 1974–1990.* New Jersey, USA: Transaction Publishers, IX.
12. Ramsbotham Oliver and Woodhouse Tom. 1999. *Encyclopedia of International Peacekeeping Operations*. Santa Barbara, California: ABC, 5.
13. Frederick H. Jr. Fleitz. 2002. *Peacekeeping Fiascos of the 1990s Causes, Solutions, and US Interests*. Wesport: Praeger, 146.
14. Ibid., 46.
15. See Uppsala Conflict Data Program (UCDP) one-sided violence dataset; US State Department and Human Rights Watch annual country reports.
16. See Security Council resolution 864 (1993).
17. See Uppsala Conflict Data Program (UCDP) one-sided violence dataset; US State Department and Human Rights Watch annual country reports.

18. See Security Council resolution SCR 976 of February 9, 1995.
19. See Uppsala Conflict Data Program (UCDP) one-sided violence dataset; US State Department and Human Rights Watch annual country reports.
20. Ramsbotham and Woodhouse 1999, 9.
21. See Security Council resolution 1118 (1997) of June 30, 1997.
22. See Uppsala Conflict Data Program (UCDP) one-sided violence dataset; US State Department and Human Rights Watch annual country reports.
23. Jett Dennis C. 2000. *Why Peacekeeping Fails.* New York: St Martins Press, 136.
24. Quoted in Ramsbotham and Woodhouse 1999, 7.
25. Ibid.
26. See "The United Nations and the Situation in Angola, May 1991–June 1994" DPI/1552/AFR/PKO,UN New York, June 1994: 3.
27. Ramsbotham and Woodhouse 1999, 8.
28. Fleitz 2002, 148.
29. George Archbald. 1998. "UN Officials Wasted Millions on Supplies: Corruption Flourished under Annan's Watch." *Washington Times*, May 29: A 1.
30. See Johannesburg Electronic Daily Mail, November 13, 1998, http://www.mg.co.za/news/98nov1/13nov-angola-html).
31. In general, see William A. Gamson. 1995. "Hiroshima, the Holocaust, and the Politics of Exclusion." *American Sociological Review*, 60: 1–20; Susan Opotow. 1995. "Drawing the Line: Social Categorization, Moral Exclusion, and the Scope of Justice." In *Conflict, Cooperation and Justice: Essays Inspired by the Work of Morton Deutsch*, ed. Barbara B. Bunker and Jeffrey Z. Rubin, San Fransisco: Josey-Bass, 347–369; Ervin Staub. 1990. "Moral Exclusion, Personal Goal Theory and Extreme Destructiveness." *Journal of Social Issues* 46: 47–64. On Rwanda, see Helen Fein. 1979. *Accounting for Genocide: National Response and Jewish Victimization During the Holocaust.* New York: Free Press Publishers; Helen Fein. 1995. "More Murder in the Middle: Life-Integrity Violations and Democracy in the World, 1987." *Human Rights Quarterly* 17: 170–191; Mark Ennals. 1988. "Ethnic Conflict Resolution and the Protection of Minorities: The Quest for NGO Competence Building." In *Ethnic Conflict and Human Rights*, ed. Kumar Rupesinghe, Tokyo: United Nations University and Norwegian University Press, 12–13; more generally, Ted R. Gurr. 1993. *Minorities at Risk: A Global View of Ethnopolitical Conflicts.* Washington, DC: U.S. The US Institute of Peace Press, 59, 82, 126.
32. Peter Uvin. 1993. "Ethnicity and Power in Rwanda and Burundi: Different Paths to Mass Violence." *Comparative Politics* 31(3), April 1999: 253–271, 255.
33. The Twa, the third demographic group in Rwanda, comprise roughly 1 percent of the population.
34. Stacey Mitchell. "The Next Genocide: Ignorance and Miscalculation in American Foreign Policy toward Rwanda." Paper presented at the annual meeting of the Midwest Political Science Association 67th Annual National Conference, The Palmer House Hilton, Chicago, IL, 2011-03-101973-94.
35. Gerard Prunier. 1995. *The Rwanda Crisis: History of a Genocide.* New York: Columbia University Press; René Lemarchand. 1970. *Rwanda and Burundi.* Westport, Connecticut: Praeger Publishers.

36. Mahmood Mamdani. 1996. *Citizen and Subject: Contemporary Africa and the Legacy of Late Colonialism*. Princeton: Princeton University Press, 24.
37. Prunier, 49.
38. A quota system was established during the First Republic that sharply delimited economic opportunities for the Tutsi (Prunier 1995, 60, 75–76).
39. Leo Kuper. 1977. *The Pity of It All*. Minneapolis: University of Minnesota Press.
40. Peter Uvin. 1999. "Ethnicity and Power in Burundi and Rwanda: Different Paths to Mass Violence." *Comparative Politics* 31(3): 253–271.
41. Alex Shoumatoff. 1992. "Rwanda's Aristocratic Guerrillas." *The New York Times Magazine*, December 13.
42. Catherine Watson. 1992. "War and Waiting." *Africa Report*, November–December, 55.
43. Report by, BBC Summary of World Broadcasts, Radio Muhabura, translated from French, November 25, 1992.
44. Charles Onyango-Obbo. 1993. "Rebellion Adds Momentum to Rwanda Reform." *Africa News Service*, April 26; David Orr. 1993. "Rwanda: Peace of a Sort Heralds End to Years of Misery." *Reuter Textline Observer*, August 1.
45. The BBTG was to be comprised of 20 ministers, with power divided roughly equally between the major political parties. In the TNA, 11 seats were awarded to each of the major parties: the Mouvement Républicain National pour la Démocratie et le Développment (MRNDD), the RPF, the Mouvement Démocratique Républican (MDR), the Parti Social Démocratie (PSD) and the Parti Libéral (PL).
46. Andre' Guichaoua, ed. 1995. *Les crises Politiques au Burundi et au Rwanda, 1993–1994: Analyses, Faits et Documents*. Lille: Universite' des Sciences et Technologies de Lille, 632–642.
47. Stacey Mitchell. "The Next Genocide: Ignorance and Miscalculation in American Foreign Policy toward Rwanda." Paper presented at the annual meeting of the Midwest Political Science Association 67th Annual National Conference, The Palmer House Hilton, Chicago, IL, 2011-03-101973-94.
48. At the time, the Tutsi comprised roughly 9–10 percent of Rwanda's population. In the BBTG the RPF was awarded 20 percent of the seats, in the TNA 16 percent. When it came to the integration of RPF forces into the Rwandan military, the distribution of power was even less equitable. Of the 13,000 troops that were to make up the new army, 40 percent were to be RPF soldiers.
49. Filip Reijntjens. 1994. *L'Afrique des Grands Lacs en crise: Rwanda, Burundi, 1988–1994*. Paris: Karthala, 117.
50. Urvin, 261, For discussions of the same processes regarding the Holocaust, see Herbert Kelman and V Lee Hamilton. 1993. "Sanctioned Massacres." In *Political Psychology: Classic and Contemporary Readings*, ed. Neil J. Kressel, New York: Paragon House, 235; John P. Sabini and Maury Silver. 1980. "Destroying the Innocent with a Clear Conscience: A Sociopsychology of the Holocaust." In *Survivors, Victims and Perpetrators: Essays on the Nazi Holocaust*, ed. Joel E. Dimsdale, Washington, DC: Hemisphere Publishing,

338–345; Daniel J. Goldhagen. 1996. *Hitler Willing Executioners: Ordinary Germans and the Holocaust*. New York: Alfred Knopf, 137. More generally, see Du Preez Wilhelmus P. 1994. *Genocide. The Psychology of Mass Murder.* London: Boyars/Bowerdean, 83, 101–107.

51. See Security Council resolution 872 (1993).
52. United Nations. 1996; Howard Adelman and Astir Suhrke. 1996. "Early Warning and Response: Why the International Community Failed to Prevent the Genocide." *Disasters* 20(4): 299.
53. Samantha Power. 2002. *A Problem from Hell: America and the Age of Genocide.* New York: Harper Perennial Publishers.
54. Prunier 1985, 231.
55. Bruce D. Jones. 2001. *The Best Laid Plans . . . Peace-Making in Rwanda and the Implication of Failure.* New York: Sage, 71–72.
56. See The Carnegie Commission on Preventing Deadly conflict, Final Report, 1994 Rwanda: UN Commander Says More troops May Have Saved Lives Inter Press Service, 191.
57. Jones 2001, 73.
58. Paul Lewis. 1994. "Boutrous-Ghali Angrily Condemns All Sides for Not Saving Rwanda." *New York Times*, May 26: A1.
59. United Nations. 1996. *The United Nations and Rwanda, 1993–1996. United Nations Blue Books series*, vol. 10. New York: United Nations, Department of Public Documents, 165.
60. Examining the effect of delayed deployment on civilian killings, the Carnegie Commission on Preventing Deadly Conflict in its final report concluded, "Early intervention by a force of 5000 could have made a significant difference in the level of violence in Rwanda."
61. UN Secretary-General Boutros Boutros-Ghali, "Report of the Secretary-General on the United Nations Assistance Mission for Rwanda," S/26927, December 30, 1993 :9–10.
62. Lansana Gberie. 2008. "Truth and Justice on Trial in Liberia." *African Affairs* 107(428): 455–465.
63. Nicole Fritz and Allison Smith. 2001. "Current Apathy for Coming Anarchy: Building the Special Court for Sierra Leone." 25 *Fordham Int'l L.J.* 391, 427.
64. Conteh-Morgan and Dixon-Fyle. 1999. *Sierra Leone at the End of the Twentieth Century: History, Politics, and Society*. Broadway New York, NY: Peter Lang International Academic Publishers, 125, 127–128.
65. Hirsch, supra note 18, 36–37; see also Bergner, supra note 18, 38 ("[T]he [diamond] mines [had] supplied [Sankoh's] army with cocaine and amphetamines, rifles and grenade launchers").
66. Hirsch, supra note 18, 32.
67. Ibid., 37–40.
68. Conteh-Morgan and Dixon-Fyle, supra note 18, 134.
69. Hirsch, supra note 18, 41–42.
70. Nicole Fritz and Allison Smith. 2001. "Current Apathy for Coming Anarchy: Building the Special." *Fordham International Law Journal* 25(2): 391–430.

71. Ibid., 64–65. Elizabeth M. Evenson. April 2004. "Truth and Justice in Sierra Leone: Coordination between Commission and Court." *Columbia Law Review* 104(3): 752.
72. Ibid., 78–79.
73. S/RES/1181 (1998) of July 13, 1998.
74. Ibid., 79.
75. K. Gallagher. 2000. "No Justice, No Peace: The Legalities and Realities of Amnesty in Sierra Leone." *Thomas Jefferson Law Review* 23: 149 et seq. (160).
76. Peace Agreement between the Government of Sierra Leone and the Revolutionary United Front of Sierra Leone, see Annex to Doc.S/1999/777(1999) of July 12, 1999.
77. Lomé Peace Accord art. XXVI(1).
78. C. Anthony. 2003. "Historical and Political Background to the Conflict in Sierra Leone." In *New Approaches in International Criminal Justice*, ed. K. Ambos/M. Othman,131 et seq. C. Anthony. 2003. "Historical and Political Background to the Conflict in Sierra Leone." In *New Approaches in International Criminal Justice, ew Approaches in International Criminal Justice: Kosovo, East Timor, Sierra Leone, and Cambodia*, ed. K. Ambos/M. Othman, Freiburg im Breisgau: Germany Max-Planck-Institutes für ausländisches und internationales Strafrecht, 131.
79. Hirsch John. 2000. *Sierra Leone: Diamonds and the Struggle for Democracy*. Boulder, Colorado: Lynne Rienner Publishers, 89.
80. RES/1313 (2000) of August 4, 2000, operative para. 3(b). Strengthening the mission had been suggested by the Secretary-General beforehand, see Fifth report of the Secretary-General on the United Nations Mission in Sierra Leone, Doc. S/2000/751.
81. Lansana Fofana. 1999. "Politics-Sierra Leone: Withdrawal Of African Peacekeepers Begins." *International Press News Agency*, http://www.ipsnews.net/1999/12/politics-sierra-leone-withdrawal-of-african-peacekeepers-begins/.
82. L. R. Hall/N. Kazemi. 2003. "Prospects for Justice and Reconciliation in Sierra Leone." *Harv. Int'l L. J.* 44: 287 et seq. (289 note 8).
83. Doc. S/2000/1091 of November 14, 2000.
84. Act No. 9, Special Agreement, 2002 (Ratification Act), 2002, reprinted in Supplement to the Gazzette Vol. CXXXIII, No. 22, at 2 (April 25, 2002) [hereinafter Special Court Ratification Act].
85. Report of the Secretary-General on the Establishment of a Special Court for sierra Leone, U.N. SC, 55th Sess., at 5, UN. Doc. S/2000/915 (2000) [hereinafter SG Report, 1].
86. Bergner, supra note 18, at 96–108.
87. Report of the Planning Mission on the Establishment of the Special Court for Sierra Leone, at 3, annex to Letter dated March 6, 2002, from the Secretary-General Addressed to the President of the Security Council, U.N. SC, 57th Sess., U.N. Doc. S/2002/ 246 (2002) [hereinafter Planning Mission Report].
88. See UNSC, Third Report of the Secretary-General on the United Nations Mission in Sierra Leone, S/2000/186, March 7, 2000, par 10.

89. Mark Malan, Rakate Phenyo and McIntyre Angela. 2002. "UNAMSIL Hits The Home Straight." *Institute for Security Studies, Monographs No. 68*, www.iss.co.za/Pubs/Monographs/No68.Chap1.html. Accessed: January 1, 2008.
90. See UNSCR resolution 1289 (2000) S/RES/1289 (2000), February 7, 2000, para 10.
91. Ibid.
92. Malan et al., 2002.
93. Ibid.
94. See United Nations Security Council resolution 1610 (2005).
95. See United Nations Mission in Sierra Leone at, www.un.org/Depts/missions/unamsil/background.html. Accessed: June, 27, 2008.
96. See Africa Report N 38 December 14, 2005.
97. Nor "North Kivu: How to end a war." ISN Security Watch. http://www.isn.ethz.ch/news/sw/details.cfm?id=18434.
98. Filip Reyntjens. 2009. *The Great African War: Congo and Regional Geopolitics, 1996–2006*. Cambridge: Cambridge UP, 45.
99. Ibid., 18.
100. Erik Kennes. 2005. "The Democratic Republic of the Congo: Structures of Greed, Networks of Need." In *Rethinking the Economics of War*, ed. Cynthia J. Arnson and I. William Zartman, Washington, D.C.: Woodrow Wilson Center, 154.
101. See "The Kivus: The Forgotten Crucible of the Congo Conflict." *Africa Report* (56), January 24, 2003.
102. See UN Security Council resolution 1291 (2000) of February 24, 2000.
103. See International Debate Report (2007), http://www.idebate.org/about/index.php.
104. Victoria K. Holt and Tobias C. Berkman. 2006. *The Impossible Mandate? Military Preparedness, the Responsibility to Protect and Modern Peace Operations*. Washington, DC: USA, 155–178.
105. See The UN-DRC Newsletter December 3, 2007.
106. V. Holt and G. Taylor. 2009. Protecting civilians in the context of UN peacekeeping operations: successes, setbacks and remaining challenges. OCHA/DPKO Jointly Commissioned Study. [August 1, 2010, Henry Stimson Center: www.stimson.org], 253–255.
107. Berkman and Holt 2006, 164–165; Holt and Taylor 2009, 259.
108. Resolution 1592, March 2005.
109. MONUC Joint Human Rights Office. (September 2009a), 2–3; Refugees International, 2010). The last line of defence: How peacekeepers can better protect civilians. [June 29, 2010, Refugees International: www.refugeesinternational.org], 7.
110. UN Security Council, October 2010, 8.
111. DR Congo calls on UN to withdraw peacekeepers, Radio Netherlands, May 18, 2011.
112. India's withdrawal of helicopters from Congo points to wider trend, Washington Post.

113. U.N. peacekeepers start Congo withdrawal, The Star, June 17, 2010.
114. "UN defends failed attempt to halt capture of Congo's Goma." Reuters. November 21, 2012. Archived from the original on November 21, 2012, http://www.reuters.com/article/2012/11/21/congo-democratic-un-idUSL5E8MLH6J20121121. Retrieved November 21, 2012.
115. See UN Security Council resolution 1291 (2000).
116. See The UN-DRC Newsletter December 3, 2007.
117. UN Office for the Co-ordination of Humanitarian Affairs (2006) MONUC Gets Tough on Foreign Armed Groups, http://209.85.165.104/search?q=cache:K0-k0nnp_7kJ:www.globalsecurity.org/military/library/news/2006/03/mil Accessed: April 24, 2008.
118. Ibid.
119. Lydia Polgreen. 2008. "Congo's Death Rate Unchanged since War Ended." *New York Times* (January 23): A8.
120. See Congressional Research Service The Library of Congress Report (2003) on Africa's Great Lakes Region: Current Conditions in Burundi, Democratic Republic of the Congo, Rwanda, and Uganda.
121. See Oxfam Report 2007, http://www.oxfamamerica.org/newsandpublications/press_releases/archive2007/oxfam-report-without-strong-un-peacekeeping-presence-congo-stands-to-lose (retrieved on April 24, 2008).
122. For more information read David Axe's article "Why South Asia Loves Peacekeeping," that appeared in The Diplomat of December 20, 2010. [January 10, 2010, The Diplomat: http://the-diplomat.com/2010/12/20/why-south-asia-loves-peacekeeping/].
123. J. Marks. 2007. "The Pitfalls of Action and Inaction: Civilian Protection in MONUC's Peacekeeping Operations." *African Security Review* 16(3): 67. Institute for Security Studies, www.issafrica.org. See also Berkman and Holt 2006, 164; Holt and Taylor 2009, 256–258; Marks 2007, 74.
124. Human Rights Watch. (December 2008). Killings in Kiwanja: The UN's inability to protect civilians. [February 9, 2009, HRW: www.hrw.org].
125. See Congressional Research Service (2003).
126. Julie Reynaert. 2011. *MONUC/MONUSCO and Civilian Protection in the Kivus*, International Peace Information Service. *file:///C:/Users/k-nsia-pepra/Downloads/20110301_monuc_JReynaert.pdf*.
127. Global Policy Forum (2008) reports that in October 2003, a UN panel of experts released a report accusing Rwanda, Uganda and Zimbabwe of systematically exploiting Congolese resources and recommended the Security Council impose sanctions at http://globalpolicy.igc.org/security/issues/kongidx.htm (retrieved on April 24, 2008).
128. Ibid.
129. *UN Expert Panel's Report 2001*.
130. Human Rights Watch. (December 2008). Killings in Kiwanja: The UN's Inability to Protect Civilians. [February 9, 2009, HRW: www.hrw.org].
131. Marks 2007, 67. Institute for Security Studies, www.issafrica.org]. See also Berkman and Holt 2006, 164; Holt and Taylor 2009, 256–258; Marks 2007, 74.
132. Dale Stephens 2005. "The Lawful Use of Force by Peacekeeping Forces: The Tactical Imperative." *International Peacekeeping* 12(2) Summer: 157.

133. Reynaert, 24.
134. Ibid.
135. Ibid., 15; See Berkman and Holt 2006, 160–161; Holt and Taylor 2009, 250–252; Reyntjens 2009, 324–325.
136. Ibid.
137. Ibid., 16.
138. Berkman and Holt 2006, 159–160; Marks 2007, 70–72.
139. Berkman and Holt 2006, 165–167; Holt and Taylor 2009, 271; Marks 2007, 75–76.
140. Reynaert, 17; Marks 2007, 77; Terrie 2009, 24.
141. K. Allen. 2007. "Congo Peacekeepers Stuck in Middle." (October 26). http://news.bbc.co.uk/2/hi/africa/7064497.*stm.*
142. Reynaert, 16.
143. Ibid., 23.
144. A recent example illustrates this. On August 18, 2010, three Indian peacekeepers were killed and six were wounded in Kirumba, near a COB of the Indian battalion. During debriefing discussions, the Indian Force Commander Chander Prakash typically sent away his other international staff officers to talk "in private" with the involved Indian peacekeepers.
145. MONUC, 31/12/2009; see Reynaert 19.
146. See Global Policy Forum Report (2008) at http://globalpolicy.igc.org/security/issues/kongidx.htm (retrieved April 24, 2008).
147. The Security Council condemned the criminal acts and the Secretary-General declared a "zero tolerance" policy. A number of ad hoc investigations were conducted, leading to the establishment of the Office for Addressing Sexual Exploitation and Abuse (OASEA) in 2005, which was transformed into the Conduct and Discipline Unit (CDU) in 2007. [May 2010, MONUSCO: www.monusco.unmissions.org].
148. For more information read the article "Casques blues aidaient les FDLR" by La Libre Belgique (July 18, 2007).
149. Pointers: Congo-Kinshasa: Under Cover, Africa Confidential, May 9, 2008, 49(10), 12; See Reynaert, 24.
150. James Munyaneza. 2007. Reports that the UN's Office for Internal Oversight Services [OIOS], New York-based internal watchdog, was investigating the matter. The office was also probing accusations that MONUC members also provide food to the rebels in return for go http://globalpolicy.igc.org/security/issues/congo/2007/0715spying.htm (retrieved on April 24, 2008).

Chapter 7

1. Tony Blair, Speech given to the Labour Party Conference, Brighton, U.K., October 2, 2001.
2. UN Chief's Rwanda Genocide Regret," *BBC News World Edition*, March 26, 2004; available at news. bbc.co.uk/2/hi/africa/3573229.stm.
3. Jemera Rone, 2005 "Sudan: The Crisis in Darfur" *Washington Post*, June 16.

4. Greg Austin and Ben Koppelman. 2004. *Darfur and Genocide: Mechanisms for Rapid Response, an End to Impunity*. London: Foreign Policy Centre.
5. The two UN reports most critical of the government of Sudan's complicity in the atrocities in Darfur are Ramcharan (2004) and Jahangir (2004). Ramcharan and Bertrand. 2004. *Report of the High Commissioner for Human Rights: Situation of Human Rights in the Darfur Region of the Sudan*, E/CN.4/2005/3, May 7; available at http://www.unhchr.ch/pdf/chr60/chr2005-3.doc; Asma Jahangir. 2004. *Report of the Special Rapporteur, Ms. Asma Jahangir*, E/CN.4/2005/7Add.2, August 6; available at http://www.ohchr.org/english/issues/executions/.

   Greg Austin and Ben Koppelman. 2004. *Darfur and Genocide: Mechanisms for Rapid Response, an End to Impunity*. London: Foreign Policy Centre.
6. Kofi Annan, speech, "Secretary-General Observes International Day of Reflection on 1994 Genocide." Geneva, April 7, 2004, http://www.unog.ch/news2/documents/newsen/sg04003e.htm.
7. Article 38(1) ICJ.
8. Cassese Antonio, 2005, *International Law*, 2nd Ed. Oxford, UK: Oxford University Press, 199.
9. Ibid.
10. Article 53 of The Vienna Convention on the Law of Treaties of 1969 (and 1986).
11. UN Conference on the Law of Treaties, First Session (1968), *Official Records* cited in A. Cassese, 203.
12. Cassese Antonio, 2005, *International Law*, 2nd Ed. Oxford, UK: Oxford University Press, 203.
13. Ibid.
14. The concept was recognized in the International Court of Justice's decision in the Barcelona Traction case [(Belgium v Spain) (Second Phase) ICJ Rep 1970 3 at paragraph 33]: See Cassesse, 16.
15. Article 2 of 1948 Genocide Conventionhttp://www.oas.org/dil/1948_Convention_on_the_Prevention_and_Punishment_of_the_Crime_of_Geno cide.pdf.
16. Ibid.
17. Eric A. Heinz. 2009. *Waging Humanitarian War: The Ethics, Law and Politics of Humanitarian Intervention*. Albany: SUNY Press, 87.
18. Annan Kofi, Secretary-General, Reflects on 35th Annual Ditchley Foundation Lecture, England, Press Release SG/SM/6613, June 26,1998.
19. See Lyal S. Sunga. Individual Responsibility in International Law for Serious Human Rights Violations, Nijhoff (1992) 252, p. IS.
20. Article 4 of the African Union's Constitutive Act.
21. In his 1998 report on the causes of conflict and possibilities for sustainable peace in Africa [. . .] (S/1998/318-A/52/871,11).
22. S/RES 1270.
23. UNSC Resolution 1270 (1999).
24. Mendez's 2001.
25. Brahimi Report 2000.

26. Zareen Iqbal. April 29, 2010. "Democratic Republic of Congo (DRC): MONUC's Impending Withdrawal". *International Institute for Justice and Development*, Retrieved January 10, 2012.
27. "Mission Statement". United Nations: Office of the special adviser on the prevention of genocide. Retrieved January 7, 2012.
28. Human Rights Watch. 2004. World Report, http://hrw.org/english/docs/2004/01/21/sudan6982.htm.
29. Samar Attar. 2010. *Debunking the Myths of Colonization: The Arabs and Europe*. Lanham, Maryland, USA: University Press of America, 252.
30. For more details, see Amnesty International 2004a; Human Rights Watch 2004a, b; International Crisis Group 2004.
31. BBC Report by Karen Allen June 18, 2007, see African action talking points on how to stop genocide in Darfur, February 13, 2007.
32. Ibid.
33. Ibid.
34. UN Resolution 1556.
35. UN Resolution 1593.
36. BBC News Report April 18, 2007.
37. International Crisis Group. 2006. *Darfur Needs Robust Protection Force.* New York, May 25.
38. Associated Press. June 25, 2007.
39. Jemera Rone, 2005 "Sudan: The Crisis in Darfur" *Washington Post*, June 16.
40. Greg Austin and Ben Koppelman. 2004. *Darfur and Genocide: Mechanisms for Rapid Response, an End to Impunity*. London: Foreign Policy Centre.
41. The two UN reports most critical of the government of Sudan's complicity in the atrocities in Darfur are Ramcharan (2004) and Jahangir (2004). Ramcharan and Bertrand. 2004. *Report of the High Commissioner for Human Rights: Situation of Human Rights in the Darfur Region of the Sudan*, E/CN.4/2005/3, May 7; available at http://www.unhchr.ch/pdf/chr60/chr2005-3.doc; Asma Jahangir. 2004. *Report of the Special Rapporteur, Ms. Asma Jahangir*, E/CN.4/2005/7Add.2, August 6; available at http://www.ohchr.org/english/issues/executions/.

    Greg Austin and Ben Koppelman. 2004. *Darfur and Genocide: Mechanisms for Rapid Response, an End to Impunity*. London: Foreign Policy Centre.
42. S/PV.5015, July 30, 2004,10–11.
43. Darfur Timeline, The UN Responds to the Crisis in Darfur: A Timeline, http://www.un.org/news/dh/dev/scripts/darfur_formatted.htm.
44. Paul D. Williams and Alex J. Bellamy. 2005. "The Responsibility to Protect and the Crisis in Darfur." *Security Dialogue* 36(1): 38.
45. Roberta Cohen and William O'Neill. 2006. "Last Stand in Sudan?" *Bulletin of the Atomic Scientists* 61(2): 52.
46. UNSC Resolution 1564.
47. UNSCR S/RES 1591, 1593, 1627, 1651, 1663, 1665, 1672, 1679.
48. S/RES 1672.
49. S/RES 1591.
50. United Nations Security Council Resolution 1593, March 31, 2005.

51. *ICC-OTP-20070227-206-E.*
52. Michael Clough. 2004. "Whose Responsibility to Protect?" *Human Rights Watch*, Volume 65 page 5. http://hrw.org/wr2k5/darfur/index.htm.
53. Cristina G. Badescu and Linnea Bergholm. 2009. "The Responsibility to Protect and the Conflict of Darfur: The Big Let Down." *Security Dialogue* 40(3): 299.
54. Ibid., 300, Resolution 1706 on Sudan was passed under Chapter VII of the UN Charter. It referred to R2P by recalling resolution 1674 on the protection of civilians in armed conflict, which had reaffirmed the R2P paragraphs of the 2005 World Summit Outcome Document.
55. UNNAMED PUBLISHER "Countdown to Sudan Carnage." *Scotland on Sunday*, September 3, 2006. http://www.scotsman.com/news/world/countdown-to-sudan-carnage-1-1414897.
56. Fischer Martin. 2008. "Morality vs Legality? Toward a Norm for the Protection of Civilians: Evidence from Darfur." *2008 Annual Conference of the International Studies Association*, 16.
57. *Reuters*. February 26, 2007. Speaking at a press conference in Addis Ababa, Ethiopia, al-Bashir reiterated his rejection of the Security Council resolution, which calls for the deployment of some 22,500 U.N. peacekeepers and police to take over the African Union mission in Darfur.
58. "Sudan Rejects 'colonial' Troops", *BBC*, June 20, 2006.
59. "UN Faces Threat over Darfur plan", *BBC*, June 10, 2006, Sudan VP vows resistance to UN peacekeepers', *AFP*, September 1, 2006.
60. Fischer 3.
61. Secretary-General's Press Conference, New York, September 13, 2006.
62. United Nations. 2007b. "Report of the Secretary-General and the Chairperson of the AU Commission on the Hybrid Operation in Darfur." S/2007/307 Rev.1, June 5.
63. Cristina and Bergholm, 300.
64. Ibid.
65. Ibid., 301.
66. Ademola Abass. 2007. "The United Nations, the African Union and the Darfur Crisis: Of Apology and Utopia." *Netherlands International Law Review* 54: 434–435.
67. Thomas Withington. 2008. *Grounded: The International Community's Betrayal of UNAMID*, a joint NGO report endorsed by a consortium of NGOs page 6; available at http://darfur.3cdn. net/b5b2056f1398299ffe_x9m6bt7cu.pdf (accessed February 1, 2009).
68. Opheera McDoom, 2008, "Darfur attacks increase local doubts over U.N." *Reuters*, February 22, http://www.reuters.com/article/2008/02/22/idUSMCD 234414.
69. Booth, 26.
70. Ibid.
71. S/PV.5153, March 29, 2005; S/RES 1591(2005), March 29, 2005.
72. S/PV.5015, July 30, 2004, 3.
73. Cristina and Bergholm, 303.

74. Bush George W. 2000. "Interview with Sam Donaldson." *ABC News This Week*, January 23.
75. Lee Feinstein and Erica De Bruin. 2009. "Beyond Words: US Policy and the Responsibility to Protect" In (eds) Richard H. Cooper, Juliette Voïnov Kohler, 2009, *Responsibility to Protect: The Global Moral Compact for the 21st Century*, New York, USA: Palgrave Macmillan, 179–195, 183.
76. Ibid., 184.
77. Scott Straus. July 2006. "Rwanda and Darfur: A Comparative Analysis." *Genocide Studies and Prevention* 1(1): 41–56,50.
78. Samuel Totten and Eric Markusen. 2005. "The US Government Darfur Genocide Investigation." *Journal of Genocide Research* 7: 279–290.
79. Scott Straus. 2005. "Darfur and Genocide Debate." *Foreign Affairs* 84: 123–133.
80. Straus, 50.
81. "Remarks by the President in Address to the United Nations General Assembly," September 21, 2004 *184. In July 2004, Congress determined that the atrocities; it urged the President to recognize the same, to pursue sanctions against the Sudanese ldrship, and to spt the establishment of a multinational peacekeeping force for the region.*
82. Colin Powell. 2004. "Powell Rules out Foreign Troops for Darfur, Urges Greater Pressure." *Sudan Sudan Tribune*, May 5, http://www.sudantribune.com/spip.php?article2801 DARFUR: RWANDA REVISITED? by Wynde Priddy DARFUR: RWANDA REVISITED? http://ww4report.com/static/darfur2.html.
83. Guy Dinmore. 2005. "White House is Quiet as Darfur Killings Continue." *Financial Times*, March 14.
84. Charles W. Corey. 2006. "United States Has Spent $1 Billion Feeding the Hungry in Darfur." *USINFO*, May 12.
85. Roberta Cohen and William O'Neill. 2006. "Last Stand in Sudan?" *Bulletin of the Atomic Scientists* 61(2): 52.
86. Cristina and Bergholm, 299.
87. Ibid., 300, Resolution 1706 on Sudan was passed under Chapter VII of the UN Charter. It referred to R2P by recalling Resolution 1674 on the protection of civilians in armed conflicts, which had reaffirmed the R2P paragraphs of the 2005 World Summit Outcome Document.
88. UNNAMED PUBLISHER "Countdown to Sudan Carnage." *Scotland on Sunday*, September 3, 2006. http://www.scotsman.com/news/world/countdown-to-sudan-carnage-1-1414897
89. United Nations. 2007b. "Report of the Secretary-General and the Chairperson of the AU Commission on the Hybrid Operation in Darfur", S/2007/307 Rev.1, June 5.
90. Cristina and Bergholm, 300.
91. The Responsibility to Protect: Report of the International Commission on Intervention and State Sovereignty.
92. Williams and Alex J. Bellamy. 2005. "The Responsibility to Protect and the Crisis in Darfur." *Security Dialogue*, 36(1): 36–40.

93. Williams and Bellamy, 38.
94. Ibid.
95. Ibid., 37.
96. Ken Silverstein. 2005. "Official Pariah Sudan Valuable to America's War on Terrorism." *The Los Angeles Times*, April 29.
97. Williams and Bellamy, 36.
98. 2004b. "Sudan Has Failed To Disarm Militias or Prevent More Attacks in Darfur—UN Envoy." *United Nations News*, October 5: 13.
99. Williams and Bellamy, 36.
100. Ramesh Thakur. 2004. "Western Medicine is no Cure for Darfur." *Australian Financial Review*, August 31, 2004a. "Sudan Can't Wait." *The Economist*, July 31: 11.
101. David McElroy. 2004. "US Forces Hunt Down Al-Qa'eda in Sudan." *Daily Telegraph*, August 1.
102. *Colgan, Ann-Louise 2006,* "A Tale of two genocides: The Failed US responses to Rwanda and Darfur," *Peacework* 369 (October) http://www.peacework magazine.org/print/309.
103. Article 4(h) of AU's.
104. Human Rights News. 2004. "Darfur: Rwandan Troops to Protect Civilians." *Human Rights Watch*, New York, August 17; available at hrw.org/english/docs/2004/08/17/darfur9241.htm.
105. Alex J. Bellamy, 2006, "Responsibility to Protect or Trojan Horse? The Crisis in Darfur and Humanitarian Intervention after Iraq" *Issue Ethics & International Affair* Volume 19, Issue 2, pages 31–54.
106. Human Rights News. 2004. "Darfur: African Union Must Insist on More Troops." *Human Rights Watch*, New York, August 20; available at hrw.org/english/ docs/2004/08/20/darfur9251.htm.
107. African Union Peace and Security Council Communiqué PSC/PR/Comm. (XVII), Peace and Security Council, 17th meeting, October 20, 2004, Addis Ababa, Ethiopia, para. 6, p. 2; available at www.africaunion.org/News_Events/Communiqu%C3%A9s/Communiqu%C3%A9%20_Eng%2020%20oct%202004.pdf.
108. "Thousands More Troops for Darfur." *BBC News* U.K. Edition, October 1, 2004; available at news.bbc. co.uk/1/hi/world/africa/3706340.stm.
109. Communiqué of 51st Meeting Peace and Security Council, May 15, 2006, Addis Ababa, Ethiopia PSC/MIN.Comm/1(LI).
110. Communiqué of the 58th Meeting of the Peace and Security Council, Banjul, The Gambia, PSC/MIN/Comm (LVIII), June 27, 2006.
111. Alhaji M. S. Bah and Ian Johnstone. 2007. "Peacekeeping in Sudan: The Dynamics of Protection, Partnerships and Inclusive Politics." *CIC Policy Paper*, May. For the details of UNAMIDs mandate see the *Report of the Secretary-General and the Chairperson of the African Union Commission on the hybrid operation in Darfur*, UN doc. S/2007/307/Rev.1, June 5, 2007. As with most contemporary UN peace operations, UNAMID is mandated to protect civilians.

112. *Communiqué* of the 45th Meeting of the Peace and Security Council, Addis Ababa, Ethiopia PSC/PR/Comm. (XLV), January 12, 2006.
113. *Communiqué* of the 46th Meeting Peace and Security Council, Addis Ababa, Ethiopia, PSC/ MIN/Comm. (XLVI), March 10, 2006.
114. Dilemmas of Regional Peacemaking: The Dynamics of the AU's Response to Darfur Dr. A. Sarjoh Bah MAR 2010 Center on International Cooperation New York University, 17.
115. Statement by the President of the Security Council, S/PRST/2006/05, February 3, 2006.
116. Dilemmas of Regional Peacemaking: The Dynamics of the AU's Response to Darfur Dr. A. Sarjoh Bah MAR 2010 Center on International Cooperation New York University, 17.
117. Ibid., 300, Resolution 1706 on Sudan was passed under Chapter VII of the UN Charter. It referred to R2P by recalling resolution 1674 on the protection of civilians in armed conflict, which had reaffirmed the R2P paragraphs of the 2005 World Summit Outcome Document.
118. Resolution 1706.
119. Fischer 16.
120. Resolution 1679 (2006), Adopted by the Security Council at its 5439th Meeting, S/RES/1679 (2006), May 06, 2006.
121. *Reuters* February 26, 2007, Speaking at a press conference in Addis Ababa, Ethiopia, al- Bashir reiterated his rejection of the Security Council resolution, which calls for the deployment of some 22,500 U.N. peacekeepers and police to take over the African Union mission in Darfur.
122. United Nations. 2007b. "Report of the Secretary-General and the Chairperson of the AU Commission on the Hybrid Operation in Darfur." S/2007/307 Rev.1, June 5.
123. African Union. 2004. "Final Communiqué: African Mini-Summit on Darfur." *Tripoli*, October 17.
124. Kristiana Powell and Stephen Baranyi. 2005. "Delivering on the Responsibility to Protect in Africa." *Policy Brief* number 4, *North South Institute*, 1–8 http://www.nsi-ins.ca/wp-content/uploads/2012/10/2005-Delivering-on-the-Responsibility-to-Protect-in-Africa.pdf.
125. Keith Gottschalk and Siegmar Schmidt. 2004. "The African Union and the New Partnership for Africa's Development: Strong Institutions for Weak States?" *Internationale Politik und Gesellschaft* 4: 138–158, 140.
126. Warrant of Arrest for Omar Al Bashir, ICC-02/05-01/09-1, March 4, 2009.
127. Security Council Resolution 1593 (2005), SC/8351, March 31, 2005.
128. Peace and Security Council Communiqué, July 21, 2008. PSC/MIN.Comm (CXLII).
129. Dilemmas of Regional Peacemaking: The Dynamics of the AU's Response to Darfur Dr. A. Sarjoh Bah MAR 2010 Center on International Cooperation New York University, 11.
130. Williams and Bellamy, 45.

131. International Crisis Group. 2006. "Getting the UN into Darfur." *Africa Briefing* 43, October 12, http://www.crisisgroup.org/home/index.cfm?id=4442&l=1.
132. International Crisis Group. 2005. The EU/AU Partnership in Darfur: Not yet a Winning Combination. Africa Report 99, October 25, http://www.crisisgroup.org/home/index.cfm?id=3766&l=1,4.
133. Keith and Schmidt 138–158, 142; International Crisis Group. 2006c. "Getting the UN into Darfur." *Africa Briefing* 43, October 12, http://www.crisisgroup.org/home/index.cfm?id=4442&l=1 (accessed March 31, 2007).

Chapter 8

1. United Nations. Report of the Panel on United Nations Peace Operations 2000.
2. Alan Bullion. 2004. "India in Sierra Leone: A Case of Muscular Peacekeeping?" *International Peacekeeping* 8(4): 77–91.
3. See Michael W. Doyle and Nicholas Sambanis. 2000. "International Peacebuilding: A Theoretical and Quantitative Analysis." *American Political Science Review* 94(4): 779, 801; Virgina Page Fortna. 2004. "Does Peacekeeping Keep Peace? International Intervention and the Duration of Peace After Civil War." *International Studies Quarterly* 48: 269, 292.
4. United Nations. 2008b. *United Nations Peacekeeping Operations Principles and Guidelines.* New York: United Nations.
5. Stian Kjeksrud. 2009. "Matching Robust Ambitions with Robust Action in Operations – Towards A Conceptual Overstretch?" (A Report). *Norwegian Defence Research Establishment* (FFI) April 20. 29
6. Gowan Richard and Benjamin Tortolani. 2008. "Robust Peacekeeping and Its Limitations." In Center on International Cooperation, *Robust Peacekeeping: The Politics of Force*, New York.
7. Victoria Holt and Glyn Taylor with Max Kelly. 2009. *Protecting Civilians in the Context of UN Peacekeeping Operations: Successes, Setbacks and Remaining Challenges, Independent Study Commissioned by DPKO and OCHA*. New York: United Nations, 4.
8. Siobhan Wills. *Protecting Civilians: The Obligation of Peacekeepers.* Oxford, UK: Oxford University Press, 82.
9. Richard Connaughton. 2001. *Military Intervention and Peacekeeping: The Reality*. London: Ashgate Publishing, 19.
10. GA Res 377 (V), 5th Sess, 302 Plen Meeting, November 3, 1950, A/PV. 302, 341at 347.
11. GAOR 5th Sess, 302st Plen Meeting, November 2, 1950, at 328.
12. *Certain Expenses of the United Nations,* Advisory Opinion of July 20, 1962, ICJ Reports 1962, 151, 163.
13. Ministry of Defense, UK, Development, Concepts and Doctrine Center, Draft UK National Doctrine that will replace JWP 3–50 The Military Contribution to Peace Support Operations, see Willis, 80.

14. Report on Wilton Park Conference 914 *Women Targetted or Afflicted by Armed Conflict: What Role for Military Peacekeepers?* May 27, 2008, 30.
15. Ibid.
16. *Report of the Independent Inquiry into the Actions of the United Nations during the 1994 Genocide in Rwanda* 26, 32.
17. Wills. Protecting Civilians: The Obligation of Peacekeepers, 71.
18. Ibid., 70.
19. Ibid.
20. W. J. Durch, ed. 2006. *Twenty-First Century Peace Operations.* Washington, DC: United Institute of Peace, 582.
21. Wills. Protecting Civilians: The Obligation of Peacekeepers, 76.
22. Volts 177.
23. Jones Bruce, Richard Gowan and Jake Sherman, 2009, "Excerpts from Building on Brahimi: Peacekeeping in an Era of Strategic Uncertainty" In *Robust Peacekeeping: The Politics of Force*, New York, US: Center on International Cooperation, New York University, 34.
24. United Nations. 2000. *The Report of the Panel on United Nations Peace Operations.* New York: United Nations A/55/305-S/2000/809.
25. Ibid.
26. Jones Bruce, Richard Gowan and Jake Sherman, 2009, "Excerpts from Building on Brahimi: Peacekeeping in an Era of Strategic Uncertainty" In *Robust Peacekeeping: The Politics of Force*, New York, US: Center on International Cooperation, New York University, 35.
27. Volts 187.
28. Report of the Panel on United Nations Peace Operations, 48–64.
29. (A/56/732 p.7).
30. George Archbald. 1998. A1, "UN Officials Wasted Millions on Supplies: Corruption Flourished under Annan's Watch." *Washington Times*, May 29: A1.
31. Guéhenno, Jean-Marie, 2009 "Robust Peacekeeping: Building Political Consensus and Strengthening Command and Control" In *Robust Peacekeeping: The Politics of Force*, New York, US: Center on International Cooperation, New York University.
32. Francis Fukuyama. 2004. *State-Building: Governance and World Order in the 21st Century.* Ithaca, NY: Cornell University Press, 92.
33. Rice Susan E. 2001. Testimony before the subcommittee on Africa of the international relations committee, United States House of Representatives, November 15.
34. Kofi Nsia-Pepra. 2014. "Militarization of US Foreign Policy in Africa Strategic Gain or Backlash?" *Military Review* January–February: 10.
35. Johan Pottier. 2002. *Re-Imagining Rwanda: Conflict, Survival and Disinformation in the Late Twentieth Century.* Cambridge: Cambridge University, 39.
36. Shaking Hands with the devil.
37. Pottier. 2002. *Re-Imagining Rwanda: Conflict, Survival and Disinformation in the Late Twentieth Century.*
38. Robust Peacekeeping: The Politics of Force, Center on International Cooperation New York University (2009) Robust Peacekeeping: Building Political

Consensus and Strengthening Command and Control Jean-Marie Guéhenno 2009, 10.

39. Ibid.
40. Ibid.
41. Dr. Bruce Jones, Richard Gowan and Jake Sherman. 2009. Building on Brahimi, Peacekeeping in an era of Strategic Uncertainty, A Report by the NYU Center on International Cooperation Submitted to the UN Departments of Peacekeeping Operations and Field Support For the "New Horizons" Project April 2009, 44.
42. Wills. Protecting Civilians: The Obligation of Peacekeepers, 28.
43. Victoria K. Holt and Glyn Taylor with Max Kelly. 2009. "Protecting civilians in the context of UN peacekeeping operations: Successes, Setbacks and Remaining Challenges" *Independent study jointly commissioned by the Department of Peacekeeping Operations and the Office for the Coordination of Humanitarian Affairs*, 7.
44. Dr. Bruce Jones, Richard Gowan and Jake Sherman, 2009, Excerpts from Building on Brahimi: Peacekeeping in an Era of Strategic Uncertainty, 23.
45. Cooper Richard H. and Juliette Voinov Kohler, 2009, "The Responsibility to Protect, The Opportunity to Relegate Atrocity Crimes to the Past" In, *Responsibility to Protect: The Global Moral Compact of the 21st Century*, Richard H Cooper and Julette Voinov Kohler eds, New York: Palgrave Macmillan.
46. Lee Feinstein and Erica De Bruin Beyond Words, US Policy and the Responsibility to Protect, 35.
47. Policy Briefno. 1/2009 Implementing the "Responsibility To protect Oxford Institute for Ethics, Law, and Armed Conflict 2009" http://www.elac.ox.ac.uk/downloads/r2p_policybrief_180209.pdf Ennifer Welsh Co-Director, Oxford Institute For Ethics, Law And Armed Conflict & Professor Of International Relations, University Of Oxford.
48. Lee Feinstein and Erica De Bruin Beyond Words, US Policy and the Responsibility to Protect, 195.
49. Ibid., 192.
50. Donald Puchala, Kate Verlin Laatikanen and Roger A. Coate. 2007. *United Nations Politics, International Organization in a Divided World, In Search of Leadership*. Upper Saddle River, New Jersey, USA: Pearson Prentice Hall.
51. Leslie H. Gelb. 2009. "The World Still Needs A Leader." *Current History* November, 108(721): 387–389, 387.
52. Ibid., 389.
53. Ibid., 388.
54. Ibid.
55. Donald et al., 91.
56. Ibid.
57. Ibid., 91.
58. Ibid., 94.
59. Ibid.

60. Time Magazine, October 18, 1999, cited in Richard Connaughton. 2001. *Military Intervention and Peacekeeping: The Reality*. London: Ashgate, 59.
61. Connaughton, 59.
62. Daily Telegraph, November 9, 1999 cited in Connaughton, 59.
63. Puchala et al., 70.
64. Leslie H. Gelb, 388.
65. Om Shanker. New York Published: November 27, 2007, http://www.nytimes.com/2007/11/27/washington/27gates.html?_r=0.
66. Puchala et al., 105.
67. Puchala et al., 111.

Bibliography

Abass, Ademola. 2007. "The United Nations, the African Union and the Darfur Crisis: Of Apology and Utopia." *Netherlands International Law Review* 54: 434–435.

Achen, Christopher H. and Duncal Snidal. 1989. "Rational Deterrence Theory and Comparative Case Studies." *World Politics* 41(2): 143–169.

Act No. 9, Special Agreement, 2002 (Ratification Act), 2002, reprinted in Supplement to the Gazzette Vol. CXXXIII, No. 22, at 2 (April 25, 2002) [hereinafter Special Court Ratification Act].

Africa Report, 38, December 14, 2005.

African Union Peace and Security Council, *Communiqué*, PSC/PR/Comm. (XVII), Peace and Security Council, 17th meeting, October 20, 2004, Addis Ababa, Ethiopia, para. 6: 2 www.africaunion.org/News_Events/Communiqu%C3%A9s/Communiqu%C3%A9%20_Eng%2020%20oct%202004.pdf.

African Union Peace and Security Council, *Communiqué* of the 45th Meeting of the Peace and Security Council, Addis Ababa, Ethiopia PSC/PR/Comm. (XLV), January 12, 2006.

African Union Peace and Security Council, *Communiqué* of the 46th Meeting Peace and Security Council, Addis Ababa, Ethiopia, PSC/ MIN/Comm. (XLVI), March 10, 2006.

African Union Peace and Security Council, *Communiqué* of 51st Meeting Peace and Security Council, May 15, 2006, Addis Ababa, Ethiopia PSC/MIN.Comm/1(LI

African Union Peace and Security Council, *Communiqué* of the 58th Meeting of the Peace and Security Council, Banjul, The Gambia, PSC/MIN/Comm (LVIII), June 27, 2006.

African Union Peace and Security Council *Communiqué*, July 21, 2008. PSC/MIN.Comm (CXLII).

Agence France Press "Sudan VP Vows Resistance to UN Peacekeepers." *AFP*, September 1, 2006.

Alao, Abiodun. 1994. *Brothers at War: Dissidence and Rebellion in Southern Africa*. London: British Academic Press.

Alexander, George L. 1979. "Case Studies and Theory Development: The Method of Structured, Focused Comparison." In *Diplomacy New Approaches in History, Theory, and Policy*, ed. Lauren Paul Gordon, New York: The Free Press.

Allen, K. 2007. "Congo Peacekeepers Stuck in Middle." *BBC News*, October 26, http://news.bbc.co.uk/2/hi/africa/7064497.stm.

Amnesty International. 1997. "Uganda-'Breaking God's Commands': The Destruction of Childhood by the Lord's Resistance Army." September 18.

Amnesty International Report. 2003.

Annan, Kofi, Secretary-General, Reflects on 35th Annual Ditchley Foundation Lecture, England, Press Release SG/SM/6613, June 26, 1998.

Annan, Kofi. 1998. Report of the Secretary-General on the situation in Africa, April 13.

Annan, Kofi. 1999. Report of the Secretary General on the Protection of Civilians in Armed Conflict, September 8.

Annan, Kofi. 1999. "Walking in the International Tightrope." *The New York Times* (September): 19.

Annan, Kofi. 1999. "Two Concepts of Sovereignty." *The Economist* (September 18): 49–50.

Annan, Kofi, speech, 2004, "Secretary-General Observes International Day of Reflection on 1994 Genocide." Geneva, April 7, http://www.unog.ch/news2/documents/newsen/sg04003e.htm.

Annual Report of the Secretary General on the Work of the Organization, June 16, 1960–June 1961, 16th Session, 1996, UN Doc A/800, 27.

Anthony, C. 2003. "Historical and Political Background to the Conflict in Sierra Leone." In *New Approaches in International Criminal Justice, New Approaches in International Criminal Justice: Kosovo, East Timor, Sierra Leone, and Cambodia*, ed. K. Ambos/M. Othman, Freiburg im Breisgau: Germany Max-Planck-Institutes für ausländisches und internationales Strafrecht.

Antonio, Cassasese. 1990. *Human Rights Achievements in a Changing World.* Philadelphia, USA: Temple University Press.

Antonio, Cassese. 2005, *International Law*, 2nd Ed. Oxford, UK: Oxford University Press.

Archbald, George. 1998. "UN Officials Wasted Millions on Supplies: Corruption Flourished under Annan's Watch." *Washington Times*, May 29: A 1.

Article 1(2) of the UN Charter.

Article 2 of the 1948 Genocide Convention http://www.oas.org/dil/1948_Convention_on_the_Prevention_and_Punishment_of_the_Crime_of_Genocide.pdf.

Article 28 of the Universal Declaration of Human Rights http://www.ichrp.org/en/article_28_udhr.

Article 39 of the UN Charter.

Article 4 of the African Union's Constitutive Act.

Assis, Malaquias. 1996. "The UN in Mozambique and Angola: Lessons Learned." *International Peacekeeping* 3: 87–103.

Austin, Greg and Ben Koppelman. 2004. *Darfur and Genocide: Mechanisms for Rapid Response, an End to Impunity*. London: Foreign Policy Centre.

Axe, David. 2010. "Why South Asia Loves Peacekeeping." *The Diplomat*, December 20, http://the-diplomat.com/2010/12/20/why-south-asia-loves-peacekeeping/.

Badescu, Cristina G. and Linnea Bergholm. 2009. "The Responsibility to Protect and the Conflict of Darfur: The Big Let Down." *Security Dialogue* 40(3): 287–309.

Bah, Alhaji M. S. and Ian Johnstone. 2007. "Peacekeeping in Sudan: The Dynamics of Protection, Partnerships and Inclusive Politics." *CIC Policy Paper*, May.

Bangkok Declaration 1967, http://www.aseansec.org/1212.htm.

BBC, Summary of World Broadcasts, Radio Muhabura, translated from French, November 25, 1992.

BBC, "Thousands More Troops for Darfur." *BBC News*, U.K. Edition, October 1, 2004; news.bbc. co.uk/1/hi/world/africa/3706340.stm.

BBC, "UN Faces Threat over Darfur plan." *BBC*, June 10, 2006.

BBC, "Sudan Rejects 'colonial' Troops." *BBC*, June 20, 2006.

Bercovitch, Jacob. 1997. "Mediation in International Conflict: An Overview of Theory, a Review of Practice." In *Peacekeeping in International Conflict: Methods and Techniques*, ed. I. W. Zartman and J. L. Rasmussen, Washington, D.C. United States: Institute of Peace.

Berdal, Mats. 2003. "Ten Years of International Peacekeeping." *International Peacekeeping* 10(4): 5–11.

Blair, Tony. 2001. *Speech given to the Labour Party Conference*. Brighton, U.K, October 2.

Boutros, Boutros-Ghali. 1992. *An Agenda for Peace*. New York: United Nations Department of Public Information.

Bowring, J. ed. 1843. *The Collected Works of Jeremy Bentham*. London.

Brahimi Report 2000. *United Nations, Report of the Panel on United Nations Peace Operations* A/55/305.

Bratt, Duane. 1996. "Assessing the Success of UN peacekeeping Operations." *International Peacekeeping* 3(4): 66–83.

Bratt, Duane. 1997. "Explaining Peacekeeping Performance: The UN in Internal Conflicts." *International Peacekeeping* 4(3): 45–70.

Briel, J. L. 1925. "Matters of Domestic Jurisdiction." *British Yearbook of International Law* VI: 8.

Brownlie, J. 1999. *Basic Documents in International Law*. Oxford: Laredon.

Bueno de Mesquita, Bruce and David Lalman. 1992. *War and Reason. Domestic and International Imperatives*. New Haven, CT: Yale University Press.

Bullion, Alan. 2004. "India in Sierra Leone: A Case of Muscular Peacekeeping?" *International Peacekeeping* 8(4): 77–91.

Burk, James. 2000. "What Justifies Peacekeeping?" *Peace Review* 12(3): 467–473.

Bush, George W. 2000. "Interview with Sam Donaldson." *ABC News This Week*, January 23.

Chantal de Jonge Qudraat. 1996. "The United Nations and Internal Conflict." In *The International Dimensions of Internal Conflict*, ed. Brown E. Michael, Cambridge, MA: MIT Press.

Claude Jr, Inis L. 1956. *Swords into Plowshare: The Problem and Progress of International Organization*. New York: Random House.

Clough, Michael. 2004. "Whose Responsibility to Protect?" 65 *Human Rights Watch*, http://hrw.org/wr2k5/darfur/index.htm.

Cohen, Roberta and William O'Neill. 2006. "Last Stand in Sudan?" *Bulletin of the Atomic Scientists* 61(2).

Cold War Canada, and Suez Crisis, file://E:/Suez Crisis.com.

Colgan, Ann-Louise. 2006. "A Tale of two genocides: The Failed US responses to Rwanda and Darfur." *Peacework* 369 (October), http://www.peaceworkmagazine.org/print/309.

Collier, Paul, Anke Hoffler and Mans Soderbom. 2004. "On The Duration of Civil War." *Journal of Peace Research* 41(3): 253–273.

Congressional Record-Senate, 139(133), October 5, 1993, p. S13043 and 139(134), October 6, 1993, p. S13146.

Coning, C. D., C. Aoiand and R. Thakur. 2007. *Unintended Consequences of Peacekeeping Operations.* New York, US: United Nations University Press.

Conlon, Justin. 2004. "Sovereignty vs. Human Rights or Sovereignty and Human Rights?" *Race and Class* 46(1): 76–100.

Connaughton, Richard. 2001. *Military Intervention and Peacekeeping: The Reality.* London: Ashgate Publishing.

Conteh-Morgan, Earl and Dixon-Fyle. 1999. *Sierra Leone at the End of the Twentieth Century: History, Politics, and Society.* Broadway New York, NY: Peter Lang International Academic Publishers.

Cooper, Richard H. and Juliette Voinov Kohler. 2009, "The Responsibility to Protect, The Opportunity to Relegate Atrocity Crimes to the Past." In *Responsibility to Protect: The Global Moral Compact of the 21st Century* ed. Richard H Cooper and Julette Voinov Kohler, New York: Palgrave Macmillan.

Corey, Charles W. 2006. "United States Has Spent $1 Billion Feeding the Hungry in Darfur." *USINFO*, May 12.

Cox, Roland. 1994. *Power and Profits.* Lexington, KY: University of Kentucky Press.

Cranston, Maurice. 1964. *What Are Human Rights*? New York: Basic Books.

Cranston, Maurice. 1967. "Human Rights: Real and Supposed." In *Political Theory and the Rights of Man*, ed. D. D. Raphael, Bloomington: Indiana University Press, 43–51.

Cranston, Maurice. 1973. *What Are Human Rights?* New York: Taplinger Publishing.

Curzon. Leslie Basil. 1998. *Dictionary of Law*, 5th Ed. London: Pitman.

Dale, Stephens. 2005. "The Lawful Use of Force by Peacekeeping Forces: The Tactical Imperative." *International Peacekeeping* 12(2), Summer: 157–172.

Dallaire, Romeo. 1995. "Towards a Rapid Reaction Capability for the United Nations." *Canadian Department of Foreign Affairs*, Ottawa, September.

Dallaire, Romeo . 2003. *Shaking Hands with the Devil: The Failure of Humanity in Rwanda.* Canada: Random House.

Dallaire, Roméo and Samantha Power. 2004. *Shake Hands with the Devil: The Failure of Humanity in Rwanda.* Cambridge, US: Da Capo Press.

Daniel, Donald C. F. and Brad C. Hayes. 1996. "Securing Observance of UN Mandates through the Employment of Military Force." *International Peacekeeping* 3(4): 105–125.

Darfur Timeline, The UN Responds to the Crisis in Darfur: A Timeline, http://www.un.org/news/dh/dev/scripts/darfur_formatted.htm.

Davenport, Christian. 1999. "Human Rights and the Democratic Proposition." *Journal of Conflict Resolution* 43(1): 92–116.

Declaration of the Rights of Man and of the Citizen [1789], http://www.britannica.com/bps/additionalcontent/8/116846/Declaration-of-the-Rights-of-Man-and-of-the-Citizen.

Diehl, Paul. 1988. "Peacekeeping Operations and the Quest for Peace." *Political Science Quarterly* 103(Autumn): 485–507.

Diehl, P. F. 1993. *International Peacekeeping.* Baltimore, US: The Johns Hopkins University Press.

Diehl, P. F. 1994. *International Peacekeeping.* Baltimore, US: The Johns Hopkins University Press.

Diehl, Paul F., Jennifer Reifschneider and Paul R. Hensel. 1996. "United Nations Intervention and Recurring Conflict." *International Organization* 50(4): 683–700.

Dinmore, Guy. 2005. "White House is Quiet as Darfur Killings Continue." *Financial Times*, March 14.

Dominick, Donald. 2003. "Neutral is Not Impartial: The Confusing Legacy of Traditional Peace Operations Thinking." *Armed Forces and Society* 29: 415.

Donnelly, Jack. 2002. *Universal Human Rights.* Ithaca, US: Cornell University Press.

Downs, George W. and Stephen John Stedman. 2002. "Evaluation Issues in Peace Implementation." In *Ending Civil Wars: The Implementation of Peace Agreements*, ed. Stephen John Stedman, Donald Rothchild and Elizabeth M. Cousens, Boulder, CO: Lynne Rienner.

Doyle, Michael W. and Nicholas Sambanis. 2000. "International Peacebuilding: A Theoretical and Quantitative Analysis." *American Political Science Review* 94(4): 779–801.

Druckman, Daniel, Stern Paul C., Diehl P., Fetherston Betts A., Johansen Robert, Durch William and Ratner Steven. 1997. "Evaluating Peacekeeping?" *Mershon International Studies Review* 41(1): 151–165.

Du Preez, Wilhelmus P. 1994. *Genocide. The Psychology of Mass Murder.* London: Boyars/Bowerdean.

Dubey, A. 2002. "Domestic Institutions and the Duration of Civil Wars Settlement." Paper presented at the Annual Meeting ISA, New Orleans.

Duffey, T. 2000. "Cultural Issues in Contemporary Peacekeeping." *International Peacekeeping* 7(1): 142–168.

Durch, William J. 1993. *The Evolution of UN Peacekeeping: Case Study and Comparative Analysis.* New York: St. Martins Press.

Durch, J. W. ed. 2006. *Twenty-First Century Peace Operations.* Washington, DC, USA: United States Institute of Peace.

Edmund, Burke and L. G. Mitchell. 2009. *Reflections on the Revolution in France.* New York: Oxford University Press.

Ennals, Mark. 1988. "Ethnic Conflict Resolution and the Protection of Minorities: The Quest for NGO Competence Building." In *Ethnic Conflict and Human Rights*, ed. Kumar Rupesinghe, Tokyo: United Nations University and Norwegian University Press.

Evenson, Elizabeth M. April 2004. "Truth and Justice in Sierra Leone: Coordination between Commission and Court." *Columbia Law Review* 104(3): 730–767.

Fearon, James D. 1995. "Rationalist Explanations for War." *International Organization* 49(3): 379–414.

Fein, Helen. 1979. *Accounting for Genocide: National Response and Jewish Victimization during the Holocaust*. New York: Free Press Publishers.

Fein, Helen. 1995. "More Murder in the Middle: Life-Integrity Violations and Democracy in the World, 1987." *Human Rights Quarterly* 17: 170–191.

Feinstein, Lee and Erica De Bruin. 2009. "Beyond Words: US Policy and the Responsibility to Protect." In (*Responsibility to Protect: The Global Moral Compact for the 21st Century*, ed. Richard H. Cooper and Juliette Voïnov Kohler, New York: Palgrave Macmillan, 179–195, 183.

Fetrerston, A. B. 1994. *Towards a Theory of United Nations Peacekeeping*. New York: St Martins' Press.

Findlay, Trevor. 2002. *The Use of Force in UN Operations*. Oxford, UK: Oxford University Press.

Finland's Parliamentary Committee for the Future.

Fischer, Martin. 2008. "Morality vs Legality? Toward a Norm for the Protection of Civilians: Evidence from Darfur." *2008 Annual Conference of the International Studies Association*, 16.

Fitfoot, P. 1992. "Functions and Powers and Inventions: UN Action in Respect of Human Rights and Humanitarian Intervention." In *To Loose the Bands of Wickedness*, ed. N. Rodley, United Kingdom: Brassey's (UK) Ltd.

Fleitz, Frederick H. Jr. 2002. *Peacekeeping Fiascos of the 1990s Causes, Solutions, and US Interests*. Wesport: Praeger.

Flinterman, Cornelius and Jeroen Gutter. 2000. *The United Nations and Human Rights: Achievements and Challenges*. Paper-UNDP Publications.

Fofana, Lansana. 1999. "Politics-Sierra Leone: Withdrawal of African Peacekeepers Begins." *International Press News Agency*, http://www.ipsnews.net/1999/12/politics-sierra-leone-withdrawal-of-african-peacekeepers-begins/.

Fortna, Virginia Page. 2003. "Inside and Out: Peacekeeping and the Duration of Peace after Civil and Interstate Wars." *International Studies Review* 5(4): 97–114.

Fortna, Virgina Page. 2004. "Does Peacekeeping Keep Peace? International Intervention and the Duration of Peace Aater Civil War." *International Studies Quarterly* 48(2): 269–292.

Fritz, Nicole and Allison Smith. 2001. "Current Apathy for Coming Anarchy: Building the Special Court for Sierra Leone." *Fordham Int'l L.J.* 25(2): 391–430.

Fukuyama, Francis. 2004. *State-Building: Governance and World Order in the 21st Century*. Ithaca, NY: Cornell University Press.

Gallagher, K. 2000. "No Justice, No Peace: The Legalities and Realities of Amnesty in Sierra Leone." *Thomas Jefferson Law Review* 23: 149 et seq. (160).

Gamson, William A. 1995. "Hiroshima, the Holocaust, and the Politics of Exclusion." *American Sociological Review*, 60: 1–20.

Gberie, Lansana. 2008. "Truth and Justice on Trial in Liberia." *African Affairs* 107(428): 455–465.

Gelb, Leslie H. 2009. "The World Still Needs A Leader." *Current History* November, 108(721): 387–389.

Gelman, Andrew and Gary King. 1994. "A Unified Method of Evaluating Electoral Systems and Redistricting Plans." *American Journal of Political Science* 38(2): 514–554.

Gene, Sharp. 1985. *Making Europe Unconquerable: The Potential Civilian-Based Deterrence and Defence.* London: Taylor and Francis.

General, Sir Michael Jackson. 2008. *The Use of Force-General Reflections,* IFS Seminar. Oslo: Institute of Defense Studies, February 5.

Ghai, Yas. 1996. "Human Rights and Governance: The Asian Debate." *International Human Rights in Context: Law, Politics, Morals,* ed. Philip Alston, Ryan Goodman and Henry J. Steiner, New York: Oxford University Press.

Gleijeses, Piero. 1988. *Conflicting Missions: Havana, Washington, and Africa, 1959–1976.* Chapel Hill, North Carolina: The University of North Carolina Press.

Goldhagen, Daniel J. 1996. *Hitler Willing Executioners: Ordinary Germans and the Holocaust.* New York: Alfred Knopf.

Goldstein, Joshua S. and Jon C. Pevehouse. 2008. *International Relations,* 8th Ed. New York: Longman.

Gottschalk, Keith and Siegmar Schmidt. 2004. "The African Union and the New Partnership for Africa's Development: Strong Institutions for Weak States?" *Internationale Politik und Gesellschaft* 4: 138–158.

Goulding, Marrack 1993. "The Evolution of United Nations Peacekeeping." *International Affairs* 69(3): 451–464.

Gowan, Richard and Benjamin Tortolani. 2008. "Robust Peacekeeping and its Limitations." In *Robust Peacekeeping: The Politics of Force,* New York, US: Center on International Cooperation, New York University.

Grono, Nick. 2006 "Briefing-Darfur: The International Community's Failure to Protect" *African Affairs* 105(421):621–631.

Guehenno, Jean Marie. 2002. "On the Challenges and Achievements of Reforming UN Peace Operations." *International Peacekeeping* 9(2): 69–80.

Guehenno, Jean Marie. 2005. The Under-Secretary General for Peacekeeping, Press Release to the Special Committee on Peacekeeping Operations 180th & 181th meetings of the General Assembly (GA/PK/183 in 2005).

Guéhenno, Jean-Marie. 2009. "Robust Peacekeeping: Building Political Consensus and Strengthening Command and Control." In *Robust Peacekeeping: The Politics of Force,* New York, US: Center on International Cooperation, New York University.

Guichaoua, André. ed. 1995. *Les crises Politiques au Burundi et au Rwanda, 1993–1994: Analyses, Faits et Documents.* Lille: Université des Sciences et Technologies de Lille, 632–642.

Gurr, Ted R. 1993. *Minorities at Risk: A Global View of Ethnopolitical Conflicts.* Washington, DC: U.S. The US Institute of Peace Press, 59, 82, 12.

Hall, L. R. and Nahal Kazemi. 2003. "Prospects for Justice and Reconciliation in Sierra Leone." *Harv. Int'l L. J.* 44: 287.

Harris, D. J. 1998. *Cases and Materials on International Law,* 5th Ed. London, UK: Sweet and Maxwell.

Hartzell, Caroline, Mathew Hoddie and Donald Rothcild. 2001. "Stabilizing the Peace after Civil War." *International Organizations* 55: 183–208.

Hass, Ernst, B. 1986. *Why We Still Need the United Nations: The Collective Management of International Conflict, 1945–1985*, Policy Papers in International Affairs. Berkley, California: Institute of International Studies, University of California.

Hegre, Havard, 2004. "Size Asymmetry, Trade, and Militarized Conflict." *Journal of Conflict Resolution* 48(3): 403–429.

Heinz, Eric A. 2009. *Waging Humanitarian War: The Ethics, Law and Politics of Humanitarian Intervention*. Albany: SUNY Press.

Hillen, John. 2000. *Blue Helmets: The Strategy of UN Military Operations*, 2nd Ed. Washington, D.C.: Brassey's.

Hirsch, John. 2000. *Sierra Leone: Diamonds and the Struggle for Democracy*. Boulder,CO: Lynne Rienner Publishers.

Holt, V. and G. Taylor. 2009. Protecting Civilians in the Context of UN Peacekeeping Operations: Successes, Setbacks and Remaining Challenges. OCHA/DPKO Jointly Commissioned Study. [August 1, 2010, Henry Stimson Center: www.stimson.org], 253–255.

Holt, Victoria K. and Tobias C. Berkman. 2006. *The Impossible Mandate? Military Preparedness, the Responsibility to Protect and Modern Peace Operations*. Washington, DC: USA, 155–178.

Holt, Victoria and Glyn Taylor with Max Kelly. 2009. *Protecting Civilians in the Context of UN Peacekeeping Operations: Successes, Setbacks and Remaining Challenges, Independent Study Commissioned by DPKO and OCHA*. New York: United Nations.

Holt, Victoria, Glyn Taylor and Max Kelly. 2009. "UN Peacekeeping Operations Successes, Setbacks and Remaining Challenges" Independent study jointly commissioned by the Department of Peacekeeping Operations and the Office for the Coordination of Humanitarian Affairs, New York: United Nations.

Howard, Adelman and Astir Suhrke. 1996. "Early Warning and Response: Why the International Community Failed to Prevent the Genocide." *Disasters* 20(4): 295–304.

Hultman, Lisa. 2006. Civilians as Military Targets: Violent Bargaining Strategies by Governments and Rebels Presented at the Annual ISA Convention, San Diego, 22–25, March.

Human Rights News. 2004. "Darfur: African Union Must Insist on More Troops." *Human Rights Watch*, New York, August 20, hrw.org/english/docs/2004/08/20/darfur9251.htm.

Human Rights News. 2004. "Darfur: Rwandan Troops to Protect Civilians." *Human Rights Watch*, New York, August 17; hrw.org/english/docs/2004/08/17/darfur9241.htm.

Human Rights Watch Country Report, http://www.hrw.org/countries.html.

Human Rights Watch. 2004. World Report, http://hrw.org/english/docs/2004/01/21/sudan6982.htm.

Human Rights Watch. 2008. " Killings in Kiwanja: The UN's inability to protect civilians." December.

Hume, David, David Fate Norton and Mary J. Norton. 2000. *Treatise of Human Nature*. New York: Oxford University Press.

Inge, Tvedten. 1997. *Angola: Struggle for Peace and Reconstruction*. Boulder, CO: Westview Press.International Commission Report 2001 titled "Responsibility to Protect" (R2P) accessed at http://www.iciss.ca/pdf/Commission-Report.pdf (August 22, 2007).

International Court of Justice Article 38(1).

International Court of Justice. 1962. *Certain Expenses of the United Nations*, Advisory Opinion of July 20, ICJ Reports 1962, 151, 163.

International Court of Justice's decision, *The Barcelona Traction case* [(Belgium v Spain) (Second Phase) ICJ Rep 1970 3.

International Crisis Group. 2005. "The EU/AU Partnership in Darfur: Not yet a Winning Combination." *Africa Report*, 99, October 25, http://www.crisisgroup.org/home/index.cfm?id=3766&l=1.

International Crisis Group. "UN Darfur Needs Robust Protection Force." May 2006.

International Crisis Group. 2006. "Getting the UN into Darfur." *Africa Briefing* 43, October 12, http://www.crisisgroup.org/home/index.cfm?id=4442&l=1.

International Debate Report. 2007. http://www.idebate.org/about/index.php.

International Relations and Security Network Report 2005.

Iqbal, Zareen. April 29, 2010. "Democratic Republic of Congo (DRC): MONUC's Impending Withdrawal." *International Institute for Justice and Development*.

Jahangir, Asma. 2004. *Report of the Special Rapporteur*, E/CN.4/2005/7Add.2, August 6, http://www.ohchr.org/english/issues/executions/.

James III, Martin W. 2011. *A Political History of the Civil War in Angola 1974–1990*. New Jersey, US: Transaction Publishers.

James, Allan. 1990. "International Peacekeeping: The Disputants' View." *Political Studies* 38(2): 215–230.

James, Art E., Gary King and Cutis S. Signorino. 2000. "Aggregation among Binary, Count, and Duration Models: Estimating the Same Quantities from Different Levels of Data." *Political Analysis* 9(1): 1–24.

Jeffrey, Frieden A. 1989. "The Economics of Intervention: American Overseas Investments and Relations with Underdeveloped Areas, 1890–1950." *Comparative Studies in Society and History* 31(1): 55–80.

Jett, Dennis C. 2000. *Why Peacekeeping Fails*. New York: St Martins Press.

Joel, Brockner. 1982. "Factors Affecting Entrapment in Escalating Conflicts." *Journal in Research in Personality* 6: 247–266.

Johansen, Robert C. 1994. "UN Peacekeeping: How Should Measure Success?" *Mershon International Studies Review* 38: 307–310.

Johnstone, Ian. 2006. "Dilemma of Robust Peace Operations." In *Robust Peacekeeping: The Politics of Force*. New York, US: Center on International Cooperation New York University.

Jones, Bruce D. 2001. *The Best Laid Plans . . . Peace-Making in Rwanda and the Implication of Failure*. New York: Sage.

Jones, Bruce, Richard Gowan and Jake Sherman, 2009. "Excerpts from Building on Brahimi: Peacekeeping in an Era of Strategic Uncertainty" In *Robust Peacekeeping: The Politics of Force*, New York, US: Center on International Cooperation, New York University.

Kalley, Jacqueline Audrey. 1999. *Southern African Political History: A Chronology of Key Political Events from Independence to Mid-1997*. Westport, CT and London: Greenwood.

Kelman, Herbert and V Lee Hamilton. 1993. "Sanctioned Massacres." In *Political Psychology: Classic and Contemporary Readings*, ed. Neil J. Kressel, New York: Paragon House.

Kennes, Erik 2005. "The Democratic Republic of the Congo: Structures of Greed, Networks of Need." In *Rethinking the Economics of War*, ed. Cynthia J. Arnson and I. William Zartman, Washington, DC: Woodrow Wilson Center, 154.

King, Gary. 1989. "Event Count Models for International Relations: Generalizations and Applications." *International Studies Quarterly* 33: 123–147.

King, Gary, Robert O Keohane and Sidney Verba. 1994. *Designing Social Inquiry: Scientific Inference in Qualitative Research*. Princeton, NJ: Princeton University Press.

Kiwanuka, Richard N. 1988. "The Meaning of 'People' in the African Charter on Human Rights and People's Rights." *82 Am. J. Int'l L.* 82(1) (January): 80–101.

Kjeksrud, Stian. 2009. "Matching Robust Ambitions with Robust Action in Operations- Towards a Conceptual Overstretch?" *Norwegian Defence Research Establishment (FFI)* April 20 (A Report).

Krause, Volker, Susumu Suzuki and Richard Witmer. 2006. "US Domestic Politics and International Cooperation, 1953–1998." *International Politics* 43: 620–643.

Kristine, Eck and Joakim Kreutz. 2006. Uppsala Conflict Data Program, One-Sided Violence Dataset 2006 version, http://www.pcr.uu.se/research/UCDP/data_and_publications/datasets.htm.

Kuper, Leo. 1977. *The Pity of It All*. Minneapolis: University of Minnesota Press.

La Libre Belgique. 2007. "Casques blues aidaient les FDLR." July 18.

Lacey, Marc. 2005. "UN Forces Using Tougher Tactics to Secure Peace." *New York Times*, May 23.

Lake, D. and D. Rothchild. 1996. "Containing Fear: The Origins and Management of Conflict." *International Security* 21(2): 41–75.

Lakhdar, Brahimi and Ahmed Salman. 2008. *Pursuit of Peace. The Seven Deadly Sins of Mediation in Robust Peacekeeping: The Politics of Force*. New York, US: Center on International Cooperation New York University.

Lemarchand, René. 1970. *Rwanda and Burundi*. Westport,CT:Praeger Publishers.

Lemke, Douglas and Jacek Kugler. 1996. "The Evolution of the Power Transition perspective." In *Parityand war. Evaluations and Extensions of the War Ledger*, ed. Jacek Kugler and Douglas Lemke, Ann Arbor: Michigan University Press.

Lewis, Paul. 1994. "Boutrous-Ghali Angrily Condemns All Sides for Not Saving Rwanda." *New York Times*, May 26: A1.

Li, Xiaorong. 1999. "Asian Values and the Universality of Human Rights." *Business and Society Review* 102/103: 81–87.

Lieberman, Evans S. 2005. "Nested Analysis as a Mixed-Method Strategy for Comparative Research." *American Political Science Review* 99(3): 435–452.

Lomé Peace Accord art. XXVI(1).

Long, Scott J. 1997. *Regression Model for Categorical and Limited Dependent Variables*. Thousand OaksUS: Sage Publishers.

Lucas, Stephen E. 1989. "Justifying America: The Declaration of Independence as a Rhetorical Document." In *American Rhetoric: Context and Criticism*, ed. Thomas W. Benson and Carbondale, Illinois: Southern Illinois University Press.

Luttwak, Edward. N. 1999. "Given War a Chance." *Foreign Affairs* 78(4): 36–44.

Mackinlay, John and Jarat Chopra. 1992. "Second Generation Multinational Operations." *The Washington Quarterly* 15(3): 113–131.

Makau, Wa Mutua. 1995. "Why Redraw the Map of Africa: A Moral and Legal Inquiry." *Mich Int'l L* 16: 1113–1176.

Makau, Wa Mutua. 1996. "The Ideology of Human Rights." *36 Virginia Journal of International Law* 589: 626–646.

Malan, Mark, Rakate Phenyo and McIntyre Angela. 2002. "UNAMSIL Hits The Home Straight." *Institute for Security Studies, Monographs No. 68*, www.iss.co.za/Pubs/Monographs/No68.Chap1.html.

Mamdani, Mahmood. 1996. *Citizen and Subject: Contemporary Africa and the Legacy of Late Colonialism*. Princeton: Princeton University Press.

Marks, J. 2007. "The Pitfalls of Action and Inaction: Civilian Protection in MONUC's Peacekeeping Operations." *African Security Review* 16(3): 67–80 Institute for Security Studies, Pretoria, South Africa, www.issafrica.org.

Mason, David and Patrick J. Fett. 1996. "How Civil War Ends: A Rational Choice Approach." *Journal of Conflict Resolution* 40(4) (December): 546–568.

Matthew, John, Letters, *The Times*, UK, November 6, 1992 (election observer).

Mbiti, John. 1990. *African Religions and Philosophy*. PortsmouthUS: Heinemann Educational Publishers.

McElroy, David. 2004. "US Forces Hunt Down Al-Qa'eda in Sudan." *Daily Telegraph*, August 1.

Messer, Ellen. 1997. "Pluralists Approaches to Human Rights." *Journal of Anthropological Research* 53(3): 293–317.

Ministry of Defense, UK, Development, Concepts and Doctrine Center, Draft UK National Doctrine that will replace JWP 3-50 The Military Contribution to Peace Support Operations.

Mitchell, Stacey. 2011. "The Next Genocide: Ignorance and Miscalculation in American Foreign Policy toward Rwanda." Paper presented at the annual meeting of the Midwest Political Science Association 67th Annual National Conference, The Palmer House Hilton, Chicago, IL, 2011–03–101973–94.

Mockaitis, Thomas R. 1999. *Peace Operations and Intrastate Conflict: The Sward or the Olive Branch?* Westport, CT: Praeger.

Mugnier, David. 2007. "North Kivu: How to End a War." *Open Democracy*. December 3, http://www.opendemocracy.net/article/north_kivu_how_to_end_a_war.

National Society for Human Rights, *Ending the Angolan Conflict*, Windhoek, Namibia, July 3, 2000 (opposition parties, massacres).

NSHR, Press Releases, September 12, 2000, May 16, 2001 (MPLA atrocities).

Nsia-Pepra, Kofi, 2011. "Robust Peacekeeping? Panacea for Human Rights Violations." *Peace and Conflict Studies* 18(2): 264.

Nsia-Pepra, Kofi. 2014. "Militarization of US Foreign Policy in Africa-Strategic Gain or Backlash?" *US Military Review* January–February.

Nyerere. Joseph 1968. *Freedom and Socialism*. London: Oxford University Press.

O'Neal, John R., Frances H. O'Neal, Zeev Maoz and Bruce M. Russett. 1996. "The Liberal Peace: Interdependence, Democracy, and International Conflict, 1950–85." *Journal of Peace Research* 33(1): 11–28.

Okere, Obinna B. 1984. "The Protection of Human Rights in Africa and the African Charter: A Comparative Analysis with the European and American Systems." *Human Rights Q* 6(2): 141–159.

Onyango-Obbo, Charles. 1993. "Rebellion Adds Momentum to Rwanda Reform." *Africa News Service*, April 26.

Opotow, Susan. 1995. "Drawing the Line: Social Categorization, Moral Exclusion, and the Scope of Justice." In *Conflict, Cooperation and Justice: Essays Inspired by the Work of Morton Deutsch*, ed. Barbara B. Bunker and Jeffrey Z. Rubin, San Fransisco: Josey-Bass, 347–369.

Organization for Security and Co-operation in Europe, http://www.osce.org/hcnm.

Organization of African Unity. International Panel of Eminent Personalities to Investigate the 1994 Genocide in Rwanda and the Surrounding Events, 2000, *Rwanda: The Preventable Genocide*, Addis Ababa: IPEP.

Orr, David. 1993. "Rwanda: Peace of a Sort Heralds End to Years of Misery." *Reuter Textline Observer*, August 1.

Paragraph 1, First Articles Common to ICCPR and ICESCR.

Parker, Nealin J. 2009. *Robust Peacekeeping: The Politics of Force.* New York, US: Center on International Cooperation New York University

Paust, Jordan J. M. Cherif Bassiouni, Michael P. Scharf, Jimmy Gurule, Leila Sadat, B. Williams and Sharon A. Williams. 2001. *Human Rights Module: Crimes against Humanity, Genocide, other Crimes against Human Rights, and War Crimes.* DurhamUS: Corilina Academic Press.

Payam, Akhavan. 1993. "Punishing War Crimes in the Former Yugoslavia: A Critical Juncture for the New World." *HRQ* 15(2): 262–289.

Peace Agreement between the Government of Sierra Leone and the Revolutionary United Front of Sierra Leone, see Annex to Doc.S/1999/777(1999) of July 12, 1999.

Philip, Alston, Ryan Goodman and Henry J. Steiner. 1996. *International Human Rights in Context: Law, Politics, Morals.* New York: Oxford University Press.

Pointers: Congo-Kinshasa: Under Cover, *Africa Confidential* May 9, 2008, 49(10): 12.

Polgreen, Lydia. 2008. "Congo's Death Rate Unchanged since War Ended." *New York Times* (January 23): A8.

Policy Brief no. 1/2009 Implementing the "Responsibility To protect Oxford Institute for Ethics", Law, and Armed Conflict 2009.

Pollis, Adamantia and Peter Schwab. 1979. "Human Rights: A Western Construct with Limited Application." In *Human Rights: Cultural and Ideological Perspectives*, ed. Adamantia Pollis and Peter Schwab, New York: Praeger Publications, 1–17.

Pottier, Johan. 2002. *Re-Imagining Rwanda: Conflict, Survival and Disinformation in the Late Twentieth Century.* Cambridge: Cambridge University.

Powell, Colin. 2004. "Powell Rules out Foreign Troops for Darfur, Urges Greater Pressure." *Sudan Tribune*, May 5.

Powell, Kristiana and Stephen Baranyi. 2005. "Delivering on the Responsibility to Protect in Africa." *Policy Brief* number 4, *North South Institute*, 1–8 http://www.nsi-ins.ca/wp-content/uploads/2012/10/2005-Delivering-on-the-Responsibility-to-Protect-in-Africa.pdf.

Power, Samantha. 2002. *A Problem from Hell: America and the Age of Genocide.* New York: Harper Perennial Publishers.

Preamble ICCPR, http://www2.ohchr.org/english/law/ccpr.htm.

Preamble, United Nations Charter, http://www.un.org/en/documents/charter.

Proclamation of Tehran, 1968, http://untreaty.un.org/cod/avl/ha/fatchr/fatchr.html.

Prunier, Gerard. 1995. *The Rwanda Crisis: History of a Genocide.* New York: Columbia University Press.

Puchala, Donald, Kate Verlin Laatikanen and Roger A. Coate. 2007. *United Nations Politics, International Organization in a Divided World, In Search of Leadership.* Upper Saddle River, NJ: Pearson Prentice Hall.

Radio Netherlands, DR Congo calls on UN to withdraw peacekeepers, May 18, 2011.

Ramcharan, Bertrand. 2004. *Report of the High Commissioner for Human Rights: Situation of Human Rights in the Darfur Region of the Sudan*, E/CN.4/2005/3, May http://www.unhchr.ch/pdf/chr60/chr2005-3.doc.

Ramesh, Thakur. 2004. "Western Medicine is no Cure for Darfur." *Australian Financial Review*, August 31.

Ramesh, Thakur and Carlyle Thayer, eds. 1995. *A Crisis of Expectations: UN Peacekeeping in the 1990s.* Boulder, CO: Westview Press.

Ramsbotham, Oliver and Woodhouse Tom. 1999. *Encyclopedia of International Peacekeeping Operations.* Santa Barbara,CA: ABC.

Reed, William. 2003. "Information, Power, and War." *American Political Science Review* 97(4): 633–641.

Refugees International. 2010. The Last Line of Defence: How Peacekeepers can Better Protect Civilians. June 29, www.refugeesinternational.org.

Regan, Patrick. 1996. "Conditions for Successful Third Party Interventions." *Journal of Conflict Resolution* 40(1): 336–359.

Regan, Patrick. 2000. Political Instability Task Force Data (State Failure) 2007.

Regan, Patrick. 2000. *Civil Wars and Foreign Powers.* Ann Arbor, MI: University of Michigan Press.

Reijntjens, Filip. 1994. *L'Afrique des Grands Lacs en crise: Rwanda, Burundi, 1988–1994.* Paris: Karthala, 1.

Report of the Planning Mission on the Establishment of the Special Court for Sierra Leone, at 3, annex to Letter dated March 6, 2002, from the Secretary-General Addressed to the President of the Security Council, U.N. SC, 57th Sess., U.N. Doc. S/2002/ 246 (2002) [hereinafter Planning Mission Report].

Report of the Secretary-General Pursuant to General Assembly Resolution 53/35: The Fall of Srebrenica A/54/549 of November 15, 1999.

Report of the Secretary-General on the Establishment of a Special Court for Sierra Leone, U.N. SC, 55th Sess., at 5, UN. Doc. S/2000/915 (2000) [hereinafter SG Report, 1].

Report of the Secretary General on Implementation of the Recommendations of the Special Committee on Peacekeeping Operations and the Panel on United Nations Operations A/55/977 June 1, 2001.

Report on Wilton Park Conference . *Women Targeted or Afflicted by Armed Conflict: What Role for Military Peacekeepers?* May 27, 2008, 30.

Reuters. 2008. "Darfur Attacks Increase Local Doubts over UN." February 22.

Reynaert, Julie. 2011. *MONUC/MONUSCO and Civilian Protection in the Kivus,* International Peace Information Service. *file:///C:/Users/k-nsia-pepra/Downloads/20110301_monuc_JReynaert.pdf*.

Reyntjens, Filip. 2009. *The Great African War: Congo and Regional Geopolitics, 1996–2006*. Cambridge: Cambridge UP.

Rice, Susan E. 2001. Testimony before the subcommittee on Africa of the international relations committee, United States House of Representatives, November 15.

Rihkye, Indar J. 1984. *The Theory and Practice of International Peacekeeping*. London, UK: Hurst and Company.

Rio Declaration on Environment and Development, 1992. http://www.unep.org/Documents.Multilingual/Default.asp?documentid=78&articleid=1163.

Roberts, Adams. 1994. "The Crisis in Peacekeeping." *Survival* 36(3):131–167.

Rone, Jemera. 2005 "Sudan: The Crisis in Darfur." *Washington Post*, June 16.

Roosevelt, Franklin D. 1944. "The Economic Bill of Rights." American Heritage Center, message to the Congress of the United States on the State of the Union http://www.fdrheritage.org/bill_of_rights.htm.

Roth, Kenneth. 2006. "UN: Darfur Needs Protection Force." *Human Rights Watch* (May 25), 18(9).

Roth, Brad R. 2004. "The Enduring Significance of State Sovereignty." *Florida Law Review* 56(5) (December): 1017–1049.

Roth, Brad. 1999. *Governmental Illegitimacy in International Law*. Oxford: Clarendon Press; New York: Oxford University Press.

Rothchild, Donald S. 1997. *Managing Ethnic Conflict in Africa: Pressures and Incentives for Cooperation*. Washington, DC: The Brookings Institution.

Rowe, Peter. 1994. "The United Nations Rules of Engagement and the British Soldier in Bosnia." *International and Comparative Law Quarterly* 43(4): 946–956.

Ruggie, John. 1993. "Wandering in the Void: Charting the UN's New Strategic Role." *Foreign Affairs* 72(4): 26–31.

Rummel. R. J. 1995. "Democracy, Power, Genocide, and Mass Murder." *Journal of Conflict Resolution* 39(1): 3–26.

Russett, Bruce and J. S. Sutterlin. 1991. "The UN in a New World Order." *Foreign Affairs* 70(2): 69–83.

Sabini, John P. and Maury Silver. 1980. "Destroying the Innocent with a Clear Conscience: A Sociopsychology of the Holocaust." In *Survivors, Victims and Perpetrators: Essays on the Nazi Holocaust*, ed. Joel E. Dimsdale, Washington, DC: Hemisphere Publishing, 338–345.

Samar, Attar. 2010. *Debunking the Myths of Colonization: The Arabs and Europe*. Lanham, MD: University Press of America.

Sarjoh, A. Bah. 2010. "Dilemmas of Regional Peacemaking: The Dynamics of the AU's Response to Darfur." *Center on International Cooperation*, New York University.

Schwartz, Eric P. 2005, " UN Peacekeeping Reform: Seeking Greater Accountability and Integrity." *Council on Foreign Affairs* http://www.cfr.org/world/un-peacekeeping-reform-seeking-greater-accountability-integrity/p8113.

Scotland on Sunday. "Countdown to Sudan Carnage." *Scotland on Sunday*, September 3, 2006. http://www.scotsman.com/news/world/countdown-to-sudan-carnage-1-1414897.

Sen, Amartya. 1997. "Human Rights and Asian Values." *The New Republic*, July 14–21.

Sexual violence. http://www.un.org/womenwatch/daw/public/cover.htm.

Shamar, Boas Elron and Ben-Ari Eyal. 1999. "Why Don't They Fight Each Other? Cultural Diversity and Operational Unity in Multinational Forces." *Armed Forces and Society* 26(1): 73–97.

Shashi, Thardoor. 1995. "Should United Nations Peacekeeping Go 'Back to Basics'?" *Survival* 37(2): 52–64.

Sherry, G. L. 1986. "The United Nations, International Conflict, and American Security." *Political Science Quarterly* 101(5): 753–771.

Shestack, Jerome J. 1998. "The Philosophical Foundations of Human Rights," *HRQ* 20(2): 201–234.

Shoumatoff, Alex. 1992. "Rwanda's Aristocratic Guerrillas." *The New York Times Magazine*, December 13.

Shue, Henry. 1980. *Basic Rights: Subsistence, Affluence, and U.S. Foreign Policy*. Princeton: Princeton University Press.

Siekmann, Robert. 1991. *National Contingents in United Nations Peacekeeping Forces*. Dordrecht: Nijhoff.

Silverstein, Ken. 2005. "Official Pariah Sudan Valuable to America's War on Terrorism." *The Los Angeles Times*, April 29.

Sim, Susan. 1996. "Human Rights: The East Asian Challenge." *The Straights Times*, Singapore, February 18.

Singer, David J. and Melvin Small. 1994. *Correlates of War Project: International and Civil War Data, 1816–1992*, ICPSR Study # 9905. Ann Arbor, MI.

Sitowski. A. 2006. *UN Peacekeeping Myth and Reality*. London and Westport: Praeger Security International.

Snyder, J. L. 2000. *From Voting to Violence: Democratization and Nationalist Conflict*. New York, USA: W. W. Norton and Company.

Soyinka, Wole. 2000. "A Fate Worse than Imperialism." *The Guardian*, Saturday review, June 3.

Staub, Ervin 1990. "Moral Exclusion, Personal Goal Theory and Extreme Destructiveness." *Journal of Social Issues* 46: 47–64.

Steiner, Henry J. and Philip Alston. 1996. *International Human Rights in Context-Law, Politics and Morals*. Calderon: Oxford University Press.

Stephen, Edwards. 2003. "The Failure of the Security Council a Blow to Chrétien Government." *National Post*, Wednesday, March 16, A9.

Stephens, Dale. 2005. "The Lawful Use of Force by Peacekeeping Forces: The Tactical Imperative." *International Peacekeeping* 12(2) Summer: 157–172.

Stockholm Declaration of the United Nations Conference on the Human Environment, 1972 http://www.unep.org/Documents.Multilingual/Default.asp?documentid=97&articleid=1503.

Straus, Scott. 2005. "Darfur and Genocide Debate." *Foreign Affairs* 84: 123–133.

Straus, Scott. July 2006. "Rwanda and Darfur: A Comparative Analysis." *Genocide Studies and Prevention* 1(1): 41–56, 50.

Stuart, Griffin. 2001. "Peacekeeping, the United Nations, and the Future Role of Commonwealth." *Commonwealth and Comparative Politics* 39(3): 150–164.

Sunga, Lyal S. 1992. *Individual Responsibility in International Law for Serious Human Rights Violations*, Nijhoff.

Thakur, Ramesh. 1994. "From Peacekeeping to Peace Enforcement: UN Operations in Somalia." *The Journal of Modern African Studies* 32(3): 387–410.

The Carnegie Commission on Preventing Deadly conflict, Final Report, 1994. Rwanda: UN Commander Says More troops May Have Saved Lives Inter Press Service, 191.

The Declaration on the Right to Development adopted by the General Assembly of the UN on December 4, 1986, http://www.un.org/documents/ga/res/41/a41r128.htm.

The Economist, 2004, "Sudan Can't Wait." July 31: 11.

The European Social Charter, 1991. http://www.coe.int/t/dghl/monitoring/socialcharter/presentation/escrbooklet/English.pdf.

The Hungarian Parliamentary Commissioner for Future Generation http://jno.hu/en/.

"The Kivus: The Forgotten Crucible of the Congo Conflict." *Africa Report* (56), January 24, 2003.

The Responsibility to Protect, 2001, Report of the International Commission on Intervention and State Sovereignty, ICISS.

The UN-DRC Newsletter December 3, 2007.

The United Nations and the Situation in Angola, May 1991–June 1994" DPI/1552/AFR/PKO, UN New York, June 1994: 3.

The Vienna Convention on Human Rights, 1993.

The Vienna Convention on the Law of Treaties of 1969, Article 53.

The Vienna Declaration, 1993 paragraph 5.

Totten, Samuel and Eric Markusen. 2005. "The US Government Darfur Genocide Investigation." *Journal of Genocide Research* 7: 279–290.

U.N. peacekeepers start Congo withdrawal, *The Star*, June 17, 2010.

UCDP One-Sided Violence Dataset v.1.2 1989-2005 http://www.pcr.uu.se/research/UCDP/our_data1.htm.

UN Security Council resolution 872 (1993).

UN Charter, Article 2(4).

UN Charter, Article 39.

UN Charter, Article 42.

UN Charter, Article 1 http://www.un.org/en/documents/charter/chapter1.shtml.

"UN Chief's Rwanda Genocide Regret." *BBC News World Edition*, March 26, 2004; available at news. bbc.co.uk/2/hi/africa/3573229.stm.

UN DPKO and DFS. 2008. *United Nations Peacekeeping Operations, Guidelines and Principles for Peacekeeping Operations*, New York: United Nations.

UN Expert Panel's Report 2001.

UN Office for the Co-ordination of Humanitarian Affairs (2006) MONUC Gets Tough on Foreign Armed Groups, http://209.85.165.104/search?q=cache:K0-k0nnp_7kJ:www.globalsecurity.org/military/library/news/2006/03/mil.

UN Report on Integrated Mission 2006 http://64.233.167.104/search?q=cache: YYgZsbeknxYJ:ochaonline.un.org/OchaLinkClick.aspx%3Flink%3Docha% 26DocId%3D1003352+report+on+integrated+missions&hl=en&ct=clnk& cd=1&gl=us.

UN resolution, RES/1313 (2000) of August 4, 2000, operative para. 3(b).

UN Secretary-General Boutros Boutros-Ghali, "Report of the Secretary-General on the United Nations Assistance Mission for Rwanda," S/26927, December 30, 1993: 9–10.

UN Secretary-General, 2000, Fifth report of the on the United Nations Mission in Sierra Leone, Doc. S/2000/751.

UN Secretary General, fifth report on the United Nations Mission in Sierra Leone, S/2000/751 July 31, 2000, paragraph 12, http://www.un.org/Docs/sc/reports/ 2001//751e.pdf.

UN Secretary-General's Press Conference, New York, September 13, 2006.

UN Security Council Documents, S/RES/2000, August 4, 2000, paragraph 3, http://www.un.org/Docs/scres/2000/2000sc1313.htm.

UN Security Council resolution SCR 976 of February 9, 1995.

UN Security Council resolution 1118 (1997) of June 30, 1997.

UN Security Council resolution 1291 (2000).

UN Security Council resolution 1291 (2000) of February 24, 2000 [S/RES/1291 (2000)].

UN Security Council resolution 1291 (2000) of February 24, 2000.

UN, Security Council Resolution 1593 (2005), SC/8351, March 31, 2005.

UN. 2004. "Sudan Has Failed To Disarm Militias or Prevent More Attacks in Darfur-UN Envoy." *United Nations News*, October 5: 13.

UN, Report of the Secretary-General and the Chairperson of the African Union Commission on the hybrid operation in Darfur, UN doc. S/2007/307/Rev.1, June 5, 2007.

UNDPO. 1996. Lessons Learned Unit, Department of UN Peacekeeping Operations, Multidisciplinary Peacekeeping: Lessons from Recent Experience, December.

UNGA Resolution 377 (V), 5th Sess, 302 Plen Meeting, November 3, 1950, A/PV. 302, 341at 347.

United Nations DPKO and Department of Field Support's Guidelines and Principles for Peacekeeping Operations, (*Capstone Doctrine*) 2008.

United Nations General Assembly. 1999. Report of the Secretary General pursuant to General Assembly Resolution 53/55: The Fall of Srebrenica; November 15.

United Nations General Assembly. 1999. *Report of the Secretary-General to the Security Council on the Protection of Civilians in Armed Conflict* S/1999/957, September 8.

United Nations General Assembly. 2000. *We the Peoples: The Role of the United Nations in the 21st Century*, Millennium Report of the Secretary-General, GA/9704 A/54/2000 of March 27.

United Nations General Assembly. 2003, Special Committee on Political and Decolonization, 58th General Assembly, Fourth Committee tenth meeting, October 17.

United Nations Mission in Sierra Leone at, www.un.org/Depts/missions/unamsil/background.html.

United Nations Press Release SG/SM/6503.

United Nations Security Council Resolution 1296 of April 2000.

United Nations Security Council Documents, S/RES/2000, August 4, 2000, paragraph 3. http://www.un.org/Docs/scres/2000/2000sc1313.htm.

United Nations Security Council Resolution 1265 of 17 September 1999, [S/RES/1265 September 17].

United Nations Security Council Resolution 1593, March 31, 2005.

United Nations Security Council resolution 1610 (2005).

United Nations, 1999, Report of the Independent Inquiry into the Actions of the United Nations during the 1994 Rwandan Genocide, New York, United Nations, December 15.

United Nations, 58th General Assembly Meeting, Fifth Committee 19th Meeting, Press release GA/AB/359 2005.

United Nations, Department of Peacekeeping Operations, www.un.org/Depts/dpko/dpko.

United Nations. 1990. *The Blue Helmets: A Review of United Nations Peacekeeping*, 2nd Ed. New York: The United Nations Department of Public Information.

United Nations. 1996. *The United Nations and Rwanda, 1993-1996. United Nations Blue Books series*, vol. 10. New York: United Nations, Department of Public Documents.

United Nations. 2004. *High Level Panel on Threats Challenges and Change: A More Secure World: Our Responsibility.* New York: United Nations.

United Nations. 2005. World Summit Outcome, para. 92 obtained at http://www.un-ngls.org/un-summit-FINAL-DOC.pdf.

United Nations. 2007 "Report of the Secretary-General and the Chairperson of the AU Commission on the Hybrid Operation in Darfur." S/2007/307 Rev.1, June 5.

United Nations. 2007. "Report of the Secretary-General and the Chairperson of the AU Commission on the Hybrid Operation in Darfur", S/2007/307 Rev.1, June 5.

United Nations. 2008. *United Nations Peacekeeping Operations Principles and Guidelines.* New York: United Nations.

United States Department of State Annual Country Reports on Human Rights Practices, http://www.state.gov/r/pa/ei/a2z/26394.htm.

United States Department of State, *Administration Policy on Reforming Multilateral Peace Operations,* April 1994, 33 ILM 705, 706 at 809-10 [hereinafter US DEPT. OF STATE].

Universal Declaration of Human Rights, 1948 http://www.un.org/en/documents/udhr/index.shtml.

UNSC resolution 1265, 1999, Adopted by the Security Council at its 4046th meeting, on September 17, 1999 S/RES/1265 (1999).

UNSC resolution 1270, (1999), S/RES/1270 (1999), October 22, 1999.

UNSC resolution 1289 (2000) S/RES/1289 (2000), February 7, 2000.

UNSC resolution 1296 (2000) S/RES/1296 (2000), Adopted by the Security Council at its 4130th meeting, on April 19, 2000.

UNSC Resolution 1592, March 2005.

UNSC, Third Report of the Secretary-General on the United Nations Mission in Sierra Leone, S/2000/186, March 7, 2000, par 10.

UNSCR resolution 1289 (2000) S/RES/1289 (2000), February 7, 2000, para 10.

Upendra, Baxi. 1999. "Voices of the Suffering, Fragmented Universality, and the Future of Human Rights." In *The Future of International Human Rights*, ed. Burns H. Weston and Stephen P. Marks, Ardsley, NY: Transnational Publishers.

US Congressional Research Service . The Library of Congress Report (2003) on Africa's Great Lakes Region: Current Conditions in Burundi, Democratic Republic of the Congo, Rwanda, and Uganda.

Uvin, Peter. 1999. "Ethnicity and Power in Burundi and Rwanda: Different Paths to Mass Violence." *Comparative Politics* 31(3): 253–271.

Vasak, Karel, 1977. "Human Rights: A Thirty-Year Struggle: The Sustained Efforts to Give Force of Law to the Universal Declaration of Human Rights." *UNESCO Courier, Paris: United Nations Educational, Scientific, and Cultural Organization*, 30(11).

Wallensteen, Peter and Margareta Sollenberg. 2001. "Armed Conflict 1989–2000." *Journal of Peace Research* 38(5): 629–644.

Walter, A. McDougal. 1997. *Promised Crusader State: The American Encounter with the World Since 1776.* Boston: Houghton Mifflin.

Waltz, Kenneth. 1959. *Man, the State, and War: A Theoretical Analysis.* New York: Columbia University Press.

Waltz, Kenneth N. 1979. *Theory of International Politics.* Reading, MA: Addison-Wesley.

Watson, Catherine. 1992. "War and Waiting." *Africa Report*, November–December, 55.

Weiss, Thomas G., David P. Forsythe and Roger A. Coate. 2001. *The United Nations and Changing World Politics*, 3rd Ed. Boulder, CO: Westview.

Weiss, Thomas G. ed. 1995. "The United Nations and Civil Wars at the Dawn of the Twenty-First Century." In *The United nations and Civil Wars*, ed. Thomas G. Weiss, Boulder, CO: Lynne Reinner.

Wellens, Karel C. 1990. *Resolutions and Statements of the United Nations Security Council (1946–1989): A Thematic Guide.* Dordrecht, The Netherlands: Martins Nijhoff Publishers.

Weston, Burns. 1995. "Human Rights." *Encyclopedia Britannica*, Chicago, 20: 656–664.

Williams, Paul D. and Alex J. Bellamy. 2005. "The Responsibility to Protect and the Crisis in Darfur." *Security Dialogue* 36(1): 27–47.

Wills, Siobhan. 2009. *Protecting Civilians: The Obligations of Peacekeepers.* New York, USA: Oxford University Press.

Withington, Thomas. 2008. *Grounded: The International Community's Betrayal of UNAMID*, a joint NGO report endorsed by a consortium of NGOs http://darfur.3cdn. net/b5b2056f1398299ffe_x9m6bt7cu.pdf.

Wittman, Donald 1979. "How a War Ends: A Rational Model Approach." *Journal of Conflict Resolution* 23(4): 743–763.

Woodhouse, Tom. 1999. "The Gentle Hand of Peace? British Peacekeeping and Conflict Resolution in Complex Political Emergencies." *International Peacekeeping* Summer 6(2): 24–37.

Index

Printed in the United States of America